Modern Machining Processes

Modern Machining Processes

Modern Machining Processes

P C PANDEY
H S SHAN
University of Roorkee, Roorkee

McGraw Hill Education (India) Private Limited
CHENNAI

McGraw Hill Education Offices
Chennai New York St Louis San Francisco Auckland Bogotá Caracas
Kuala Lumpur Lisbon London Madrid Mexico City Milan Montreal
San Juan Santiago Singapore Sydney Tokyo Toronto

 **McGraw Hill Education (India) Private Limited**

ISBN (13 digit): 978-0-07-096553-9
ISBN (10 digit): 0-07-096553-6

Published by McGraw Hill Education (India) Private Limited, 444/1, Sri Ekambara Naicker Industrial Estate, Alapakkam, Porur, Chennai 600 116, Tamil Nadu, India, and printed at Sai Printo Pack Pvt. Ltd., New Delhi 110 020

Visit us at: www.mheducation.co.in

Foreword

It is now well established that conventional machining methods are unable to meet the challenges posed by the demand for economic machining of ultra hard and high strength materials to close tolerances. Extensive research on this subject over the last two decades has led to the development of a number of newer unconventional machining methods. Some of these processes are now well established and form a part of the standard facilities available in a modern workshop.

On account of the importance of these production methods in modern manufacturing systems, modern machining methods are now included in the curriculum of most graduate and undergraduate production engineering courses in the country and abroad. Yet, there are very few really good textbooks available on this topic. The authors must, therefore, be congratulated for this timely publication.

This book grew out of an earlier monograph on modern machining processes published by the authors in February 1976, under the auspices of the Mechanical and Industrial Engineering Curriculum Development Centre at the University of Roorkee. The authors have revised and expanded that monograph considerably and have presented a balanced account of both the theory and practice of modern machining methods. I feel sure that the book will serve as a good textbook for students of mechanical and production engineering courses both at the graduate and undergraduate levels. The book will also be useful as a reference for practising engineers and designers working in industries and research institutions.

S LAL

Director
Indian Institute of Technology
Kharagpur

Preface

The concept of material removal by an edged tool, involving plastic deformation and formation of chips, has been known to man for several hundred years. In recent years an increasing demand for the machining of components of complex shapes made of hard, difficult-to-machine materials with exacting tolerances and surface finish has resulted in the development of a number of new machining processes. Some of these methods have been commercially exploited while others are still in their experimental stages. Although a large volume of literature concerning these methods is available, most of it is scattered and has not as yet been compiled systematically.

This book has been written with a view to present a balanced account of both the theory and practice of modern machining processes and includes the results of some work done at the University of Roorkee. The material embodied in this book has been developed from lecture notes, seminars and research publications available in this area.

The authors are grateful to their colleagues for their help and suggestions in the preparation of the manuscript. Thanks are also due to Mr. Mahendra Singh and Mr. R.K. Tyagi for preparing the illustrations and Mr. Jamshed Ali for typing the manuscript.

P C PANDEY

H S SHAN

Contents

List of Symbols

q	Work hardening capacity
r	Radius of circular tool, m
t	Any instant of time
u	Velocity of a particle on concentrator, m s^{-1}
u_0	Velocity of vibrations of end face of concentrator, m s^{-1}
v	Velocity of sound in magnetostrictive material, m s^{-1}
x	Tool-work gap, or any distance, m
x_n	Coordinate of vibration node (Fig. 2.4), m
y	Displacement of tool, m
$\dot{y}$	Velocity, m s^{-1}
A	Projected area of abrasive grain, m^2
B	$\left[1-\left\{\dfrac{\beta C}{\omega}\right\}^2\right]^{-\frac{1}{2}}$
C	Velocity of sound in the concentrator material, m s^{-1}
C'	$C \cdot B$
D	Diameter of concentrator at any distance x, m
D_1	Diameter of concentrator at the tool end, m
D_0	Diameter of concentrator at the transducer end, m
E	Young's modulus of material, kg m^{-2}
$\bar{F}$	Mean static force, N
F, P	Force, N
$F(t)$	Force at any instant of time, N
G	Shear modulus of material, kg m^{-2}
H_a	Hardness of abrasive material
H_t	Hardness of tool material
H_w	Hardness of work material
$\epsilon, K_0, K_1, K_2, K_3, K_4$	Constants
K	Probability factor
K	Concentration factor, or transformation ratio
L	Length of hole, m
M	Volume ratio of abrasive and carrier gas
N	Number of abrasive particles taking cut at a time
S	Cross-sectional area of concentrator at distance x, m^2
S_1	Cross-sectional area of concentrator at the tool end, m^2
S_0	Cross-sectional area of the end of concentrator, m^2
S_r	Radiating surface area of magnetostrictor, m^2
T	Time per cycle, s
ΔT	A short interval of time, s
V, Q	Volume rate of material removed, m^3 s^{-1}
V_0	Volume of material removed per cycle, m^3
V_g	Volume of material removed per grit, m^3
V'	Volume of slurry, m^3
Z	Size of abrasive particle, m
Z_m	Maximum size of abrasive particle, m
$\bar{Z}$	Average size of abrasive particle, m

β	Half taper index
θ_z	Function to describe volume of material removed by an abrasive particle as related to the depth of penetration
λ	Wavelength, m
λ_0	$\lambda(1+\phi)$
ξ	Plastic flow stress, N m^{-2}
ρ	Density of material, kg m^{-3}
ρ_a	Density of abrasive material, kg m^{-3}
σ	Applied stress, N mm^{-2}
σ_m	Maximum applied stress, N m^{-2}
σ_y	Yield stress of material, N m^{-2}
σ	Stress on the work surface, N m^{-2}
σ'	Stress on the tool surface, N m^{-2}
ϕ	A factor
ϕ_z	Function to describe the force on an abrasive grain as related to the depth of penetration
ψ_z	Function to describe distribution of size of abrasive particles
ω	Angular velocity, $2\pi f$

CHAPTER 3

f	Tool feed rate, m s^{-1}
f_p	Maximum permissible tool feed rate, m s^{-1}
h	Tool-work gap (minimum), m
j, x, y, z	Variables
m	Mass of work material, kg
m_e	Mass of electrolyte passing through gap, kg
n	Valency of work material
q	Volume rate of flow of electrolyte, m^3 s^{-1}
s	Specific metal removal rate, m^3 amp^{-1} s^{-1}
t	Time for which current flows, s
A	Area of cross-section of tool, m^2
A_0	Area of current path, m^2
C	Specific heat of electrolyte, cal kg^{-1} K
E	Electric potential, volts
H	Heat to raise m_e kg of electrolyte from T_1 to T_b K, cal
I	Current, amp
I_p	Maximum permissible current, amp
L	Tool-work gap in the direction of tool feed, m
N	Atomic weight of work material
P	Power equivalent of H, W
R	Resistance, ohm
S	Current density, amp m^{-2}
T_b	Boiling point of electrolyte, K
T_1	Inlet temperature of the electrolyte, K

ΔT	Difference in temperature of electrolyte at inlet and exit of tool-work gap, K
V_e	Volume of electrolyte passing through the gap, m³
α, γ	Angles, deg
δ	Prefix to indicate an increment
η	Current efficiency
ρ	Density of work material kg m⁻³
ρ_e	Density of electrolyte, kg m⁻³
ρ_s	Specific resistance of electrolyte, ohm-m

CHAPTER 4

a	Area of surface where molecule reacts, m²
a_0	Area of surface of one molecule, m²
b	$1/\lambda$
d	Gap between the tool-work electrodes, m
e	Electron charge
f	Charging frequency, c/s
f_e	Electron pressure, N m⁻²
f_h	Hydrostatic pressure due to molten surface, kg m⁻²
f_s	Force due to surface tension per unit length, N m⁻¹
f'	Net heat flow rate per unit area, cal m⁻²
g	Acceleration due to gravity, m s⁻²
h	Depth of cavity or crater produced, m
h'	Constant in Eq. 4.44
i	Current density, amp m⁻²
i_c	Charging current, amp
i_d	Discharge current, amp
k	Boltzmann constant
l	Constant in Eq. 4.44
m	Exponent
m_e	Electron mass, kg
m_g	Mass of gas molecule, kg
m_1	Mass of material removed per unit area of plane surface, kg m⁻²
m_0	Mass of material removed per unit surface area per unit time, kg m⁻² s⁻¹
m_p	Mass of proton, kg
n	Exponent
n_e	Number of electrons per unit area per unit time, m⁻² s⁻¹
n_0	Number of molecules per unit area per unit time, m⁻² s⁻¹
n'	Number of atoms leaving film surface per unit area per unit time, m⁻² s⁻¹
p	Pressure of etchant, N m⁻²
p_a	Back pressure of evaporating atoms, N m⁻²
r	Radius of hole or cavity produced, m

s	Thermal conductivity of material, $W\ m^{-1}\ K^{-1}$
t	Time, s
t'	Charging time of the condenser, s
v_a	Atomic velocity, $m\ s^{-1}$
v_{ab}	Steady ablation velocity, $m\ s^{-1}$
v_e	Velocity of electrons, $m\ s^{-1}$
x	Any distance, m
A	Surface area, m^2
C	Capacitance, Farad
C_p	Specific heat of material, $J\ kg^{-1}\ K^{-1}$
E	Energy per spark, J
F_a	Anode work function
F_{bp}	Force due to back pressure, N
F_c	Cathode work function
F_e	Force due to electron bombardment, N
F_h	Force due to hydrostatic pressure, N
F_s	Force due to surface tension, N
H	Total heat to vaporize the material from room temp., $J\ kg^{-1}$
H_{cla}	Centre line average value of the surface roughness, μ
H_f	Heat of fusion, $J\ kg^{-1}$
H_v	Heat of vaporization, $J\ kg^{-1}$
I	Current, amp
$K,\ K',\ K_1,\ K_2,\ K_3,\ K_4,\ K_5,\ A,\ B,$	Constants
L	Equivalent inductance of discharge circuit, ohm
M	Atomic weight
M_T	$\dfrac{M_w}{M_t}$
M_t	Melting point of the tool electrode material, K
$M_w,\ T_{mo}$	Melting temperature of work material, K
N_0	Number of unreacted molecules on surface
N_1	Number of reacted molecules on surface
O	Over cut, m
P	Power to vaporize the material, $J\ s^{-1}$
Q	Heat contained in the work material, J
R	Resistance, ohm
R_f	Growth rate of film surface, $m^2\ s^{-1}$
R_t	Average metal removal rate from the tool electrode, $m^3\ amp^{-1}\ s^{-1}$
T	Temperature, K
T_b	Boiling temperature of material, K
T_1	Initial temperature of material, K
T_m	Melting temperature of work material, K
U	Accelerating voltage, V
U_b	Breakdown voltage, V
U_c	Condenser voltage at the instant of spark initiation, V
U_s	Supply voltage, V
V	Rate of volume of material removed, $m^3\ s^{-1}$

V_c	Cathode fall
V_1	Volume of material removed per discharge. m^3
V_m	Volume of one molecule, m^3
W	Specific energy of vaporization of material, $J\ m^{-3}$
W_R	Wear ratio (work/tool)
X	Number of bombarding atoms
X_c	Characteristic depth, as defined in Eq. 4.49
Y	Number of surface atoms taken away by chemical reaction
α	Taper angle, deg
β	Thermal diffusivity of material
η	Operating efficiency or cutting efficiency
θ, ψ	Atomic species in Eq. 4.45
λ	Mean free path of electron
ρ	Density of work material, $kg\ m^{-3}$
τ	Mean time of stay of unreacted molecule on surface, s

1
Modern Machining Processes : An Overview

There has been a rapid growth in the development of harder and difficult-to-machine metals and alloys during the last two decades. Conventional edged tool machining is uneconomical for such materials and the degree of accuracy and surface finish attainable is poor. The advancing strength level would have a catastrophic effect on the total machining bill if there was no corresponding improvement in machining technology (Fig. 1.1). In

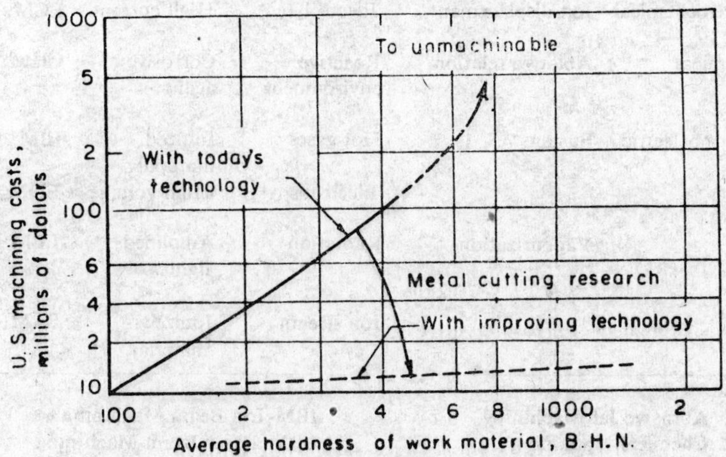

Fig. 1.1 Effect of work material strength on national machining cost in the United States of America (1)

view of the seriousness of this problem, Merchant (1960) emphasized the need for the development of newer concepts in metal machining. By adopting a unified programme and utilizing the results of basic and applied research, it has now become possible to process some of the materials which were formerly considered to be unmachinable under normal conditions. The newer machining processes, so developed, are often called 'modern machining processes' or 'unconventional machining methods'. These are unconventional in the sense that conventional tools are not employed for metal cutting; instead energy in its direct form is utilized.

Modern machining methods are classified according to the type of fundamental machining energy employed, for example, mechanical, electrochemical, chemical or thermoelectric. Table 1.1 gives a classification of the machining processes based on the type of energy used, the mechanism of metal removal, the source of energy requirements, etc.

Table 1.1

Classification of Modern Machining Processes

Type of energy	*Mechanism of metal removal*	*Transfer media*	*Energy source*	*Processes*
Mechanical	Erosion	High velocity particles	Pneumatic/ hydraulic pressure	AJM, USM, WJM
	Shear	Physical contact	Cutting tool	Conventional machining
Electrochemical	Ion displacement	Electrolyte	High current	ECM, ECG
Chemical	Ablative relation	Reactive environment	Corrosive agent	CHM
Thermoelectric	Fusion	Hot gases	Ionized material	IBM, PAM
		Electrons	High voltage	EDM
	Vapourization	Radiation	Amplified light	LBM
		Ion stream	Ionized material	PAM

NOTE

AJM	Abrasive Jet Machining	IBM	Ion Beam Machining
CHM	Chemical Machining	LBM	Laser Beam Machining
ECG	Electrochemical Grinding	PAM	Plasma Arc Machining
ECM	Electrochemical Machining	USM	Ultrasonic Machining
EDM	Electric Discharge Machining	WJM	Water Jet Machining

PROCESS SELECTION

To make efficient use of modern machining processes, it is necessary to know the exact nature of the machining problem. It is to be understood that: (i) these methods cannot replace the conventional machining processes and (ii) a particular machining method found suitable under the given conditions may not be equally efficient under other conditions. A careful selection of the process for a given machining problem is, therefore, essential.

Before selecting the process to be employed, the following aspects must be studied:

(i) Physical parameters.

(ii) Properties of the work material and the shape to be machined.

(iii) Process capability.

(iv) Economic considerations.

When comparing the physical parameters of modern machining processes (Table 1.2), it may be noticed that both EDM and USM require approximately the same power, whereas ECM consumes roughly forty times more power than EDM. In Table 1.3, it can be seen that although ECM consumes much greater power, it is an excellent method for drilling long slender holes with length/dia ratio $>$ 20.

Table 1.2

Physical Parameters of the Modern Machining Processes

Parameters	USM	AJM	ECM	CHM	EDM	EBM	LBM	PAM
Potential (V)	220	220	10	—	45	150,000	4500	100
Current (Amp)	12 (A.C.)	1.0	10,000 (D.C.)	—	50 (pulsed D.C.)	0.001 (pulsed D.C.)	2 (average 200 peak)	500 (D.C.)
Power (W)	2400	220	100,000	—	2700	150 (average 200 peak)	—	50,000
Gap (mm)	0.25	0.75	0.20	—	0.025	100	150	7.5
Medium	Abrasive in water	Abrasive in gas	Electrolyte	Liquid chemical	Liquid dielectric	Vaccum	Air	Argon or hydrogen

Table 1.3

Shape Applications of Modern Machining Processes

Process	Holes				Through cavities		Surfacing		Through cutting	
	Precision small holes		Standard				Double contouring	Surfaces of revolution		
					Precision	Standard			Shallow	Deep
	Dia <.025 mm	Dia >.025 mm	Length <20 Dia	Length >20 Dia						
USM	—	—	A	C	A	A	C	—	C	—
AJM	—	—	B	C	C	B	—	—	A	—
ECM	—	—	A	A	B	A	A	B	A	A
CHM	B	B	—	—	C	B	—	—	A	—
EDM	—	—	A	B	A	A	B	—	C	—
LBM	A	A	B	C	C	C	—	—	A	B
PAM	—	—	B	—	C	C	—	C	A	A

NOTE

A Good   B Fair   C Poor

Materials applications of the various machining methods are summarized in Table 1.4. It can be seen that for the machining of electrically non-conducting materials, both ECM and EDM are unsuitable, whereas the mechanical methods can achieve the desired results.

Table 1.4

Materials Applications

| Process | Material | | | | | | | |
	Aluminium	Steel	Super alloys	Titanium	Refractories	Plastics	Ceramics	Glass
USM	C	B	C	B	A	B	A	A
AJM	B	B	A	B	A	B	A	A
ECM	B	A	A	B	B	D	D	D
CHM	A	A	B	B	C	C	C	B
EDM	B	A	A	A	A	D	D	D
EBM	B	B	B	B	A	B	A	B
LBM	B	B	B	B	C	B	A	B
PAM	A	A	A	B	C	C	D	D

NOTE

A Good Application   B Fair   C Poor   D Not Applicable

The process capabilities of modern machining methods have been compared in Table 1.5. It is to be noted that although ECM results in

Table 1.5

Process Capability

Process	Metal removal rate mm^3/min	Tolerance μ	Surface finish CLA μ	Surface damage depth μ	Corner radii mm
USM	300	7.5	0.2- 0.5	25	0.025
AJM	0.8	50	0.5- 1.2	2.5	0.100
ECM	01500	50	0.1- 2.5	5.0	0.025
CHM	15.0	50	0.4- 2.5	50	0.125
EDM	800	15	0.2-12.5	125	0.025
EBM	1.6	25	0.4- 2.5	250	2.50
LBM	0.1	25	0.4-1.25	125	2.50
PAM	75000	125	Rough	500	—
Conventional milling of steel	50000	50	0.4- 5.0	25	0.050

Table 1.6

Effects on Equipments and Tooling

Process	Tool wear ratio	Machining medium contamination	Safety	Toxity
USM	10	B	A	A
AJM	—	B	B	A
ECM	0	C	B	A
EDM	6.6	B	B	B
EBM	—	B	B	A
LBM	—	A	B	A
PAM	—	A	A	A

Tool Wear Ratio = volume of work material removed/volume of tool electrode removed

NOTE
A No Problem   B Normal Problem   C Critical Problem

excellent surface finish, it can cause extensive surface damage as compared to AJM or USM. ECM has another advantage of a very low tool wear ratio (Table 1.6) but it has certain fairly serious problems regarding the contamination of the electrolyte used and the corrosion of machine parts.

A comparison of the process economy and their relative efficiencies are given in Table 1.7 and Fig. 1.2 respectively.

Table 1.7
Process Economy

Process	Capital investment	Tooling and fixtures	Power requirement	Efficiency	Tool consumption
USM	B	B	B	D	C
AJM	A	B	B	D	B
ECM	E	C	C	B	A
CHM	C	B	D	C	A
EDM	C	D	B	D	D
EBM	D	B	B	E	A
LBM	C	B	A	E	A
PAM	A	B	A	A	A
Conventional machining	B	B	B	A	B

NOTE

A Very Low Cost   B Low   C Medium   D High   E Very High

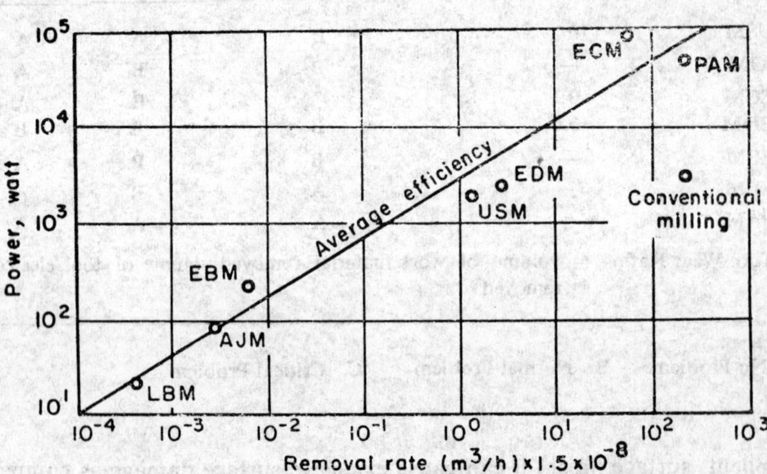

Fig. 1.2 Comparison of some modern machining processes

2
Mechanical Processes

The present trend is towards the development of material removal processes that employ primarily nonmechanical energy. 'Abrasive jet' and 'ultrasonic' machining methods are important because of certain special characteristics with which electrochemical, chemical or thermal processes cannot compete.

The mechanical processes described in the following pages, namely, ultrasonic machining, abrasive jet machining and water jet machining have been developed for commercial applications in engineering manufacture.

ULTRASONIC MACHINING

The term 'ultrasonic' is used to describe a vibratory wave of a frequency above that of the upper frequency limit of the human ear; it generally embraces all frequencies above about 16 kc/s. The upper frequency range of ultrasonics is largely dependent on the generator, practical limitations imposing a maximum frequency in the region of 500 Mc/s. The range of wavelengths in varying media is very wide. For example, when propagated in a solid medium, a wave with the frequency of 25 kc/s will have a wavelength of about 200 mm, while one with a frequency of 500 Mc/s will have a wavelength of the order of 0.008 mm.

There are two types of waves, namely, shear waves and longitudinal waves. Longitudinal waves are mostly used in ultrasonic applications since they are easily generated. They can be propagated in solids, liquids and gases and can travel at a high velocity so that their wavelength is short in most media. The velocity of the wave can be calculated for a given solid material using Eq. 2.1.

$$\text{sonic velocity} = \frac{E}{\rho}\left[\frac{m(m-1)}{(m^2-m-2)}\right]^{\frac{1}{2}} \tag{2 1}$$

where E is Young's Modulus, $1/m$ is Poisson's Ratio, and ρ is the density of the material.

The device for converting any type of energy into ultrasonic waves is called ultrasonic transducer. Most industrial applications employ transducers energized by electrical power at the required frequency. The electrical energy is converted into mechanical vibrations, and for this the piezoelectric effect in natural or synthetic crystals, or the magnetostrictive effect exhibited by some metals is utilized.

In addition to machining or cutting, ultrasonics find many other engineering applications. Some of these are:

(i) Casting and welding of metals.

(ii) Forming of plastics.

(iii) Measurement of velocity of moving fluids.

(iv) Measurement of density, viscosity and elastic constants.

(v) Measurement of hardness and grain size determination of metals.

(vi) Nondestructive residual stress determination.

(vii) Flaw detection, leak detection, etc.

The use of ultrasonics in the medical field for the diagnosis and treatment of certain diseases has also been reported.

Ultrasonic machining is a process in which material is removed due to the action of abrasive grains. The abrasive particles are driven into the work surface by a tool oscillating normal to the work surface at high frequency. Figure 2.1 illustrates the cutting action of the tool.

The tool is made of soft material, oscillated at frequencies of the order of 20 to 30 kc/s with an amplitude of about 0.02 mm. It is pressed against the work piece with a load of a few kilograms and fed downwards continuously as the cavity is cut in the work. The tool is shaped as the approximate mirror image of the configuration of the cavity desired in the work (Fig. 2.1).

The excitement to the tool is mostly given by means of a magnetostrictive transducer. A general arrangement of an ultrasonic machining head is shown in Fig. 2.1.

The slurry, which is made of abrasive particles suspended in a liquid, is fed into the cutting zone under pressure. A concentration of about 30 per cent is the optimum from the point of view of pump design and of achieving adequate penetration.

The rate of machining is not a function of load or imposed stresses on the tool. There appears to be no simple relationship between the machining rate and the physical characteristics of work material, such as tensile strength, per cent elongation, microhardness or impact strength.

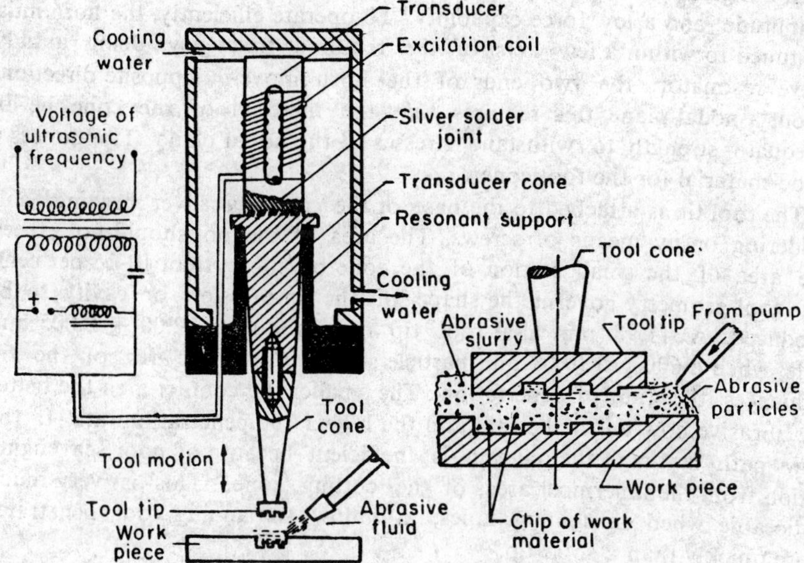

Fig. 2.1 Schematic representation of ultrasonic machining process

Elements of Process

The four main elements of this process are:

  (i) Work material.

  (ii) Tool cone and tool tip (vibration amplifier).

  (iii) Abrasive slurry.

  (iv) Ultrasonic machine.

Work Material

Earlier, it was assumed that material is removed in this process only by brittle failure, and so only brittle materials were thought to be machinable by this process. But now it has been confirmed that chips can form in this process, that is, ductile failure can also take place. There appears to be no limitation to the range of materials that can be machined, except that they should not dissolve in the slurry media or react with it. Soft and ductile materials, however, are usually cut more economically by other methods.

Tool Cone and Tool Tip

The tool cone (also called 'horn') amplifies and focuses the mechanical energy produced by the transducer and imparts this to the work piece in such a way that energy utilization is optimum. It is simply a velocity transformer with the exception that it is made slightly shorter than the half wavelength.

The horn mechanically modifies the vibratory energy to give the required force-amplitude ratio. Thus, a low gain horn gives a low amplitude with a high force capability, and conversely, a high gain horn has a high amplitude and a low force capability. To operate efficiently, the horn must be tuned to within a few kc/s of the required frequency. Being a half-wave resonator, the two ends of the horn move in opposite directions about a nodal plane; this requires that the material of the cone be of adequate strength to withstand stresses at the nodal plane. Titanium is a good material for the tool cone.

The tool tip is attached to the base of the cone by silver brazing, soft soldering or by means of screws. The area of the tool should not exceed the area of the small section of the cone by more than 10-15 per cent. The tool geometry governs the shape of the impression or cavity to be produced. A 11.98 mm dia tool tip may produce a 12.00 ± 0.005 mm hole, when a 600 grit (0.01 mm particle size) is used. The area of the tip influences the rate of penetration. The smaller the contact area the better the abrasive flow under the tool and the higher the penetration rate. If the flow path is long, the cutting is inefficient because of poor scavenging action from the innermost areas of the cutting zone. This is very much noticeable when machining holes; a hollow trepanning tool penetrates more quickly than a solid tip.

The choice of material for the tool is very vital because the cost of making the tool and the time required to change tools are critical factors in the economics of ultrasonic machining. Also, the tool tip has to withstand vibrations and it should not fail or wear out quickly. Most of the wear occurs at the end; wear at the sides is about ten times less. The use of tungsten carbide as a tool material presents many problems in shaping the tool; also the cost of such a tool will be high. Tough malleable materials, such as alloy steel and stainless steel prove satisfactory.

Abrasive Slurry
Some of the many types of abrasives in use are:
  (i) Aluminium oxide (alumina).
  (ii) Boron carbide.
  (iii) Silicon carbide.
  (iv) Diamond dust.

Boron is the most expensive abrasive material but is best suited for cutting tungsten carbide, tool steel and precious stones. Silicon finds maximum application. The problem with alumina is that it wears fast and soon loses its cutting power. Alumina is best for cutting glass, germanium and ceramics. Diamond and rubies are nicely cut by using diamond powder which ensures good accuracy, surface finish and cutting rates. Boron silico carbide is a new promising abrasive which has an abrasive power 8-12 per cent greater than that of boron carbide.

The size of abrasives varies between 200 and 2000 grit. Coarse grades are good for roughing, whereas finer grades (say 1000 grit) are used for finishing. The extremely fine grades of 1200 to 2000 grit are used only for a finishing pass over jobs of extreme accuracy. The typical surface roughness resulting from two different grain sizes are:

280 grit   0.5 micron surface roughness.
800 grit   0.2 micron surface roughness.

In actual practice, the surface roughness of the machined face is governed by the work material, roughness on tool surface, vibration amplitude, fineness of abrasive grit and efficient slurry circulation.

Liquid Media

The abrasive is suspended in liquid. The liquid performs many functions:

(i) Acts as an acoustic bond between the work piece and the vibrating tool.
(ii) Helps efficient transfer of energy between the work piece and tool.
(iii) Acts as a coolant.
(iv) Provides a medium to carry the abrasive to the cutting zone.
(v) Helps to carry away the worn abrasive and swarf.

The characteristics of a good suspension media (the liquid) are:

(i) Density, approximately equal to that of abrasive.
(ii) Good wetting properties to wet the tool, work and abrasive.
(iii) High thermal conductivity and specific heat for efficient removal of heat from the cutting zone.
(iv) Low viscosity to carry the abrasive down the sides of the hole between the tool and work piece.
(v) Noncorrosive properties to avoid corrosion of the work piece and tool.

Water is frequently used as the liquid carrier since it satisfies most of the requirements. Some inhibitor is generally added to the water.

Ultrasonic Machine

A general view of an ultrasonic machine is given in Fig. 2.2. Its main parts are:

(i) Acoustic head.
(ii) Tool feed mechanism.
(iii) Abrassive feed system.
(iv) Generator.

Acoustic head   The magnetostriction type of transducers which utilize the effect of longitudinal magnetostriction are now very common. These may be made of nickel, iron-cobalt (permendur) or iron-aluminium (alfer). Nickel finds maximum application because of high strength and good insulating properties of the nickel oxide film.

An alternating electromagnetic field of ultrasonic frequency is produced by a conventional tube generator. Under the action of this field, the magnetostrictor is periodically magnetized and its length changes. Since the deformation is independent of the direction of the electromagnetic field, the frequency of the vibrations of the magnetostriction assembly is twice that of the driving field. Premagnetization of the vibrator is done to remove this frequency doubling and to obtain maximum deformation. To

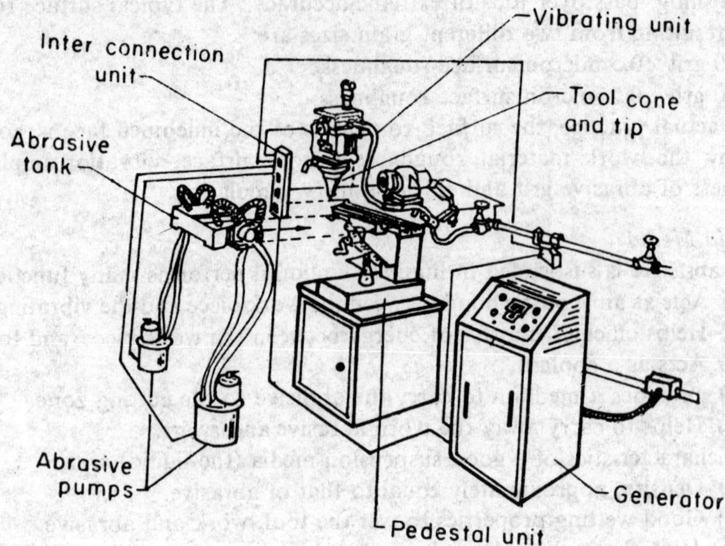

Fig. 2.2 General arrangement of an ultrasonic machine used for cutting

prevent the passage of direct current through the output transformer of the generator and the shorting of the A.C. circuit by the direct current, a choke and capacitor is used (Fig. 2.1).

A fair amount of the energy input to the transducer appears as heat; a good cooling system is, therefore, a necessity. Air cooling is used in machines up to 50 W capacity, but for high power rating machines, water cooling is a must.

Tool feed mechanism The feed mechanism of an ultrasonic machine must perform the following functions:

   (i) Bring the tool very slowly close to the work piece.
   (ii) Provide adequate cutting force and sustain this during cutting.
   (iii) Decrease the force at a specified depth.
   (iv) Overrun a small distance to ensure the required hole size at the exist.
   (v) Return the tool

For accurate working, it is vital that the feed mechanism be precise and sensitive. Figure 2.3 (a), (b), (c) and (d) show the principles of different types of feed systems. In the systems shown in Fig. 2.3 (a) and (b) counter-weights are used, the force being the difference between the weight of the head and that of the counterweight attached through a pulley or lever system. The force is adjusted through the weights. Though simple, such a system is insensitive and inconvenient to adjust. Figure 2.3 (c) shows a compact spring-loaded system which is quite sensitive. For high rating machines, pneumatic [Fig. 2.3 (d)] or hydraulic systems may be used.

Some means of reading tool displacement is often incorporated in the design of machines. The guides and other moving parts are designed to have low friction.

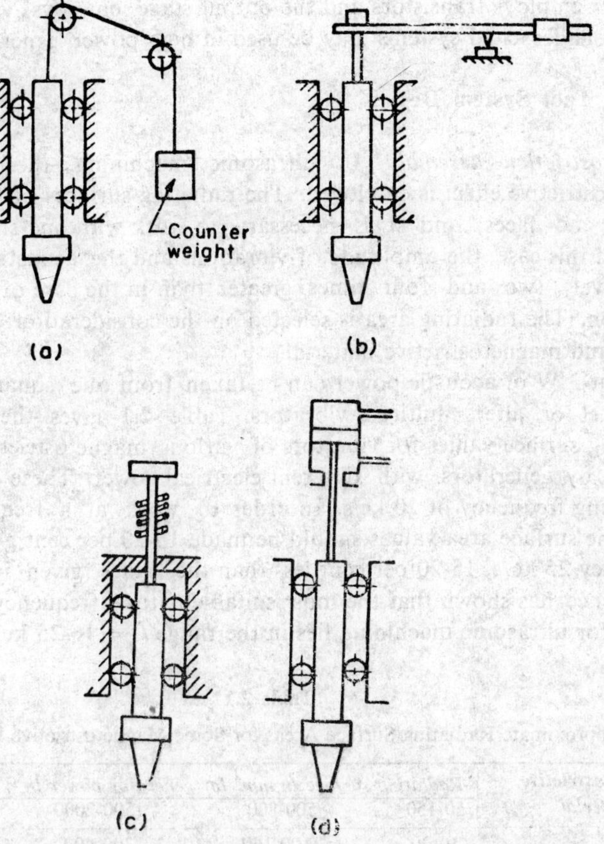

Fig. 2.3 Schematic representation of some types of tool feed
systems in ultrasonic machining

Abrasive feed system The abrasive slurry can be supplied by hand in a
small machine but for machines of higher power, a pump (usually centri-
fugal type) is used to supply the slurry through a nozzle. A good method
is to keep the slurry in a bath in the cutting zone. This ensures a good
supply and reduces any tendency of the tool to scatter the slurry when
amplitude is large.

Another effective method of supplying the slurry to the cutting zone is
via a hollow tool or holes in the work piece.

Generator The main requirements of a generator are reliability, efficiency,
simplicity in design and low cost. Vacuum tube generators are employed
usually. Small generators usually consist of a master oscillator. a buffer
amplifier and an output stage. These have a wide tuning range but low
efficiency. The master oscillator is often of the RC type. The buffer

amplifier employs transistors and the output stage employs vacuum tubes. Various half-excited systems may be used in high power generators.

Cutting Tool System Design

Magnetostriction assembly In ultrasonic machining, the longitudinal magnetostrictive effect is employed. The radiating surfaces of the assembly are its end faces, and it is necessary to work with one-sided radiation since, in this case, the amplitude of vibrations and the acoustic power are respectively two and four times greater than in the case of double-sided radiation. The radiating area is selected on the consideration of generator power and magnetostrictive material.

About 3 W of acoustic power can be taken from one square millimetre of nickel or alfer multirod vibrators. Table 2.1 gives the approximate radiating surface values for vibrators of various magnetostrictive materials excited by generators with different electrical power. These data apply to a working frequency of 20 kc/s. In order to work at a frequency of 16 kc/s, the surface area values should be made 15-20 per cent greater, and at frequency 25 kc/s, 15-20 per cent less than the values given in Table 2.1. Experience has shown that the most suitable natural frequency of vibrators meant for ultrasonic machining lies in the range $f = 16\text{-}25$ kc/s.

Table 2.1

Approximate Radiating Surface Areas for Some Magnetostrictive Materials

Magnetostrictive material	Radiating surface in mm² for generator power in W			
	50-150	500-800	1500-2000	4000-5000
Nickel	10-30	100-160	300-400	800-1000
Alfer	15-45	150-720	450-600	1200-1500
Permendur	6-18	60-96	180-240	480-600

Resonance frequency f on the main harmonic is given by

$$f = \frac{v}{\lambda} \tag{2.2}$$

where v, the velocity of sound in the material of the magnetostrictor is determined by Eq. 2.1, and λ is the wavelength.

The dimensions of a multirod vibrator can be determined from the formula

$$\tan\left\{2\pi\,\frac{l}{\lambda_0}\right\} \cdot \tan\left\{2\pi\frac{l'}{2\lambda_0}\right\} = i \tag{2.3}$$

where l is the length of the end piece and l' is the length of the rods; λ_0 is the wavelength at resonance frequency with allowance for the effect of bending of the end piece by the co-vibrating mass of the cooling liquid. $\lambda_0 = \lambda(1+\phi)$ where $\phi = 0.05$ for f less than 25 kc/s; $i = S_0/S_r$, where S_0 :

the cross section of the rod and S_r is the area of the radiating surface of the magnetostrictor.

In some instances, the use of single-rod vibrators has been recommended, but such vibrators cannot be used effectively in industrial appliances with a power of more than 250 W. On the other hand, multirod assemblies make possible a compactly designed vibrator, ensure good cooling conditions and require a much smaller number of turns in the driving and premagnetizing coils.

Tool cone (*concentrator*) When the size of magnetostriction assembly has been determined, the dimensions of the concentrator rod (velocity transformer) can be found out. A rigorous solution to the problem of the propagation of low amplitude vibrations in a rod of finite length with reducing cross section presents mathematical difficulties. Hence, several simplifying assumptions may be made for technical calculations. These are:

(i) A plane wave is propagated in the rod, that is, the stresses and velocities of the particles over the whole area of a cross section are constant.

(ii) There is no transverse compression of the rod.

(iii) The vibrations of the rod are harmonic.

Particle velocity u can be described by the following differential equation

$$\frac{d^2u}{dx^2} + \frac{1}{S} \cdot \frac{dS}{dx} \cdot \frac{du}{dx} + \frac{\omega^2}{C^2} \cdot u = 0 \tag{2.4}$$

where x is the coordinate measured from the broad end of the transformer, ω is the angular velocity ($\omega = 2\pi f$) and C is the velocity of sound in the material of the rod.

This equation holds for a cone of any shape but it is valid only where the greatest diameter of the concentrator is small in comparison with λ and if its broadening does not exceed a certain critical value. Experimental data indicate that exponential concentrators give good results in ultrasonic machining.

Exponential concentrator of circular cross section Consider the case of an exponential concentrator (Fig. 2.4) for which

$$S = S_o e^{-2\beta x}$$

where 2β is the taper index.

A general solution of Eq. 2.4 with boundary conditions taken into account takes the form

$$u = u_0 \left(\cos \frac{\omega x}{C'} - \frac{\beta C'}{\omega} \sin \frac{\omega x}{C'} \right) \exp (\beta x) \tag{2.5}$$

where u_0 is the velocity of vibrations of the end face of magnetostriction assembly and

$$C' = \cfrac{C}{\sqrt{1-\left(\dfrac{\beta C}{\omega}\right)}} = C \cdot B \qquad (2.6)$$

That is, in the concentrator, the velocity of propagation of sound waves is increased by value B. Further, velocity C' is real only when

$$\beta < \frac{\omega}{C}$$

When this inequality holds, the acoustic energy is not reflected from the lateral surface of the velocity transformer and is gradually concentrated on to an ever smaller area. The amplitudes and velocities of the vibrations correspondingly increase, as shown in Fig. 2.4 for a half-wave concentrator.

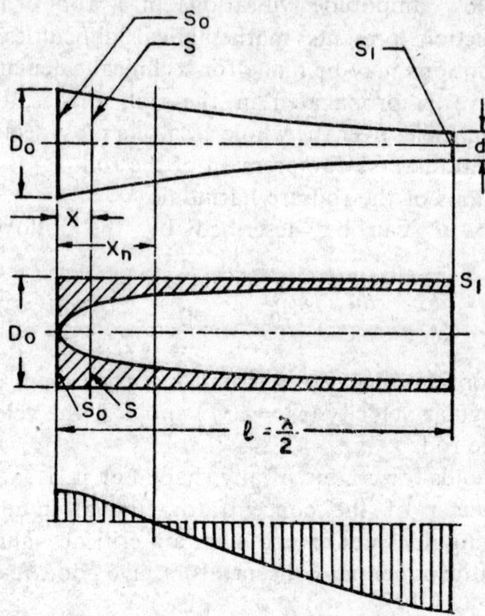

Fig. 2.4   Exponential types of tool cone used
in ultrasonic machining

Two types of exponential concentrators are used (Fig. 2.4). One is solid with a varying outer diameter and the other is hollow with a constant outer diameter but a varying inner diameter. The latter type is particularly suitable for cutting holes of large diameters right through the work piece.

If the concentration factor, (the transformation ratio) K be defined as

$$K = \frac{D_o}{d} = \sqrt{\frac{S_o}{S_1}} \qquad (2.7)$$

the taper index for an exponential concentrator can be determined from

$$\beta = \frac{\omega}{C} \cdot \frac{\ln K}{\sqrt{\pi^2 + (\ln K)^2}} \; m^{-1} \qquad (2.8)$$

The value of K should be chosen taking into consideration machining conditions and the properties of magnetostrictive materials. For example, $K = 4$ to 5 for rough through machining and $K = 3$ to 4 for finish machining. For machining hardened steel and alloys, and for nickel vibrators, higher values of K may be taken. For the cutting of glass or ceramics, however, lower values of K are chosen.

The length of concentrator l is usually chosen as half wave ($n=1$) or full wave ($n=2$)

$$l = \frac{n}{2} \cdot \lambda$$

For an exponential concentrator

$$l = \frac{nC}{f}\sqrt{1 + \left\{\frac{\ln K}{2\pi n}\right\}^2}$$

The coordinate of vibration node x_n (Fig. 2.4) can be determined by putting $l=0$ in Eq. 2.5. For a half-wave concentrator, it is seen that

$$\cot \frac{\pi x_n}{l} = \frac{1}{\pi} \cdot \ln K$$

or

$$x_n = \frac{l}{\pi} \operatorname{arc\,cot}\left(\frac{1}{\pi} \ln K\right) \qquad (2.9)$$

To summarize, the method of design of a circular half-wave exponential concentrator (with given values of f, D_0, d, C) is

$$K = \frac{D_0}{d} \; ; \; l = \frac{C}{2f}\sqrt{1 + \left\{\frac{\ln K}{\pi}\right\}^2}$$

$$\beta = \frac{\ln K}{l}$$

and from the formula

$$D = D_0 \exp^{-\beta x}$$

the shape of the concentrator can be determined. A check may be made to ensure that operating frequency f sufficiently exceeds critical frequency f_c given by

$$f_c = \frac{\beta C}{2\pi}$$

If the check does not give the necessary results, this means that too large an amplification factor K has been used, in which case it is necessary to reduce the value of K.

Exponential concentrator of rectangular cross section This type of concentrator (Fig. 2.5), which is widely used for ultrasonic cutting, can be designed as follows:

$$S_0 = a_0 \cdot b_0$$
$$S_1 = a_1 \cdot b_0$$
$$S = a \cdot b_0$$

For exponential taper

$$S = S_0 \cdot \exp^{(-2\beta x)} = a_0 \cdot b_0 \cdot \exp^{(-2\beta x)}$$

Also

$$a = a_0 \exp^{(-2\beta x)}$$

and

$$K = \sqrt{\frac{S_0}{S_1}} = \sqrt{\frac{a_0}{a_1}}$$

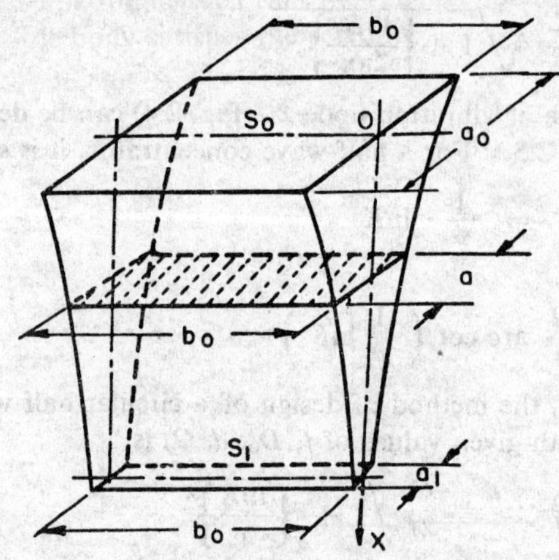

Fig. 2.5

In designing the concentrator, the values of a_0, b_0, a_1, n, f and C are first of all assigned and then the following calculations are made

$$K = \sqrt{\frac{a_0}{a_1}}$$

$$l = n \cdot \frac{C}{f} \sqrt{1 + \left\{ \frac{\ln K}{2\pi n} \right\}^2}$$

$$\beta = \frac{\ln K}{l}$$

$$a = a_0 \exp^{(-2\beta x)}$$

A final check can be made that

$$f \geqslant 1.5 \frac{\beta C}{2\pi}$$

EXAMPLE

Design a half-wave steel concentrator to work at a frequency of 19 kc/s. The transducer has a section of 7.5×7.5 cm.

Here

$$a_0 = b_0 = 75 \text{ mm}$$
$$f = 19 \text{ kc/s}$$
$$n = \tfrac{1}{2}$$

Taking the velocity of sound in steel, $C = 5 \times 10^6$ mm/sec and assuming $a_1 = 4$ mm, it is seen that

$$K = \sqrt{\frac{a_0}{a_1}} = 4.33$$

and

$$\ln K = 1.4656$$

The length of concentrator

$$l = \tfrac{1}{2} \cdot \frac{5 \times 10^6}{19 \times 10^3} \sqrt{1 + \left\{ \frac{1.4656}{\pi} \right\}^2} = 145.2 \text{ mm}$$

The exponential factor

$$\beta = \frac{\ln K}{l} = \frac{1.4656}{145.2} = 0.01009 \text{ mm}^{-1}$$

The law of change of shape is

$$a \text{ (mm)} = 75 \exp^{(-0.02018x)}$$

Where x is measured from the transducer end. Table 2.2 is prepared for thickness values at different sections along the length of the concentrator.

Table 2.2

x mm	$0.02018x$	$\exp^{(-0.02018x)}$	a mm
0	0	1	75
10	0.2018	0.81726	61.29
20	0.4036	0.66791	50.09
30	0.6054	0.54586	40.94
40	0.8072	0.44610	33.45
50	1.0090	0.36458	27.34
60	1.2108	0.29796	22.35
70	1.4126	0.24351	18.26
80	1.6144	0.19901	14.93
90	1.8162	0.16264	12.20
100	2.0180	0.13292	9.97
110	2.2198	0.10863	8.15
120	2.4216	0.08878	6.66
130	2.6234	0.07256	5.44
140	2.8252	0.05930	4.45
145.2	2.9314	0.05332	4.00

To check

$$\text{Critical frequency } f_c = \frac{\beta C}{2\pi} = \frac{0.01009 \times 5 \times 10^6}{2\pi}$$

$$= 8.032 \text{ kc/s}$$

Operating frequency $f = 19$ kc/s

As $f > 1.5 f_c$, the concentrator will work satisfactorily.

Hollow Cylindrical Concentrator

This type of concentrator (Fig. 2.6) which is used for trepanning, that is, cutting along the contour of the desired opening, can be designed as follows:

Diameter D is to be found such that the change of the cross sectional area of concentrator body satisfies the relation

$$S = S_0 \cdot \exp^{-2\beta x}$$

But

$$S = \frac{\pi}{4}(D_0^2 - D^2)$$

$$S_0 = \frac{\pi}{4} D_0^2$$

and

$$S_1 = \frac{\pi}{4}(D_0^2 - D_1^2)$$

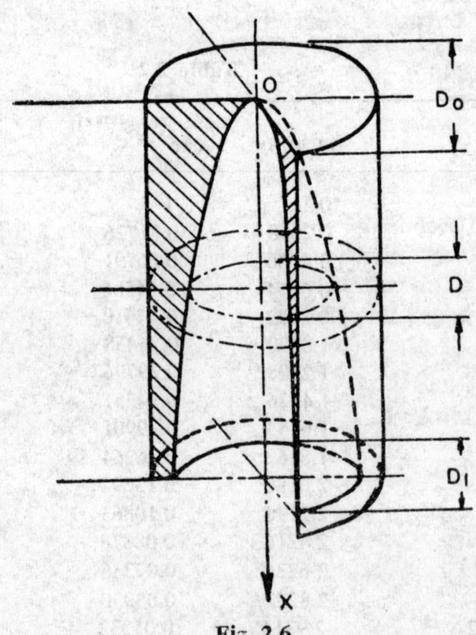

Fig. 2.6

The amplification factor

$$K = \frac{S_0}{S_1} = \sqrt{\frac{D_0^2}{(D_0^2 - D_1^2)}}$$

After simplification, it is seen that

$$D = D_0 \sqrt{1 - \exp^{-2\beta x}} \qquad (2.10)$$

In the design, therefore, the values of D_0, D_1, n, f and C are assigned and the following calculations are made

$$K = \sqrt{\frac{D_0^2}{(D_0^2 - D_1^2)}}$$

$$l = n \cdot \frac{C}{f} \sqrt{1 + \left\{\frac{\ln K}{2\pi n}\right\}^2}$$

$$\beta = \frac{\ln K}{l}$$

$$D = D_0 \sqrt{1 - \exp^{-2\beta x}}$$

A check can be made to see that operating frequency $f \geqslant 1.5$ critical frequency f_c.

EXAMPLE

Design a compound acoustic head to operate at 20 kc/s for the ultrasonic trepanning of an opening of 9 mm diameter in sheets of an aerospace material. The dimensions of the transducer surface are 4.5×4.5 cm. The change in the frequency of the transducer power supply due to tool wear is limited to 1 kc/s.

SOLUTION

In this case, the transducer will be made doubly half-wave, consisting of a half-wave continuous circular exponential member II and an exponential hollow cylindrical member III connected to an annular cylindrical tool (portion IV), as shown in Fig. 2.7. It has to be noted that lengths III and IV together constitute a half-wavelength.

By approximating the oscillating system as a whole to a cylinder of $3/2\lambda$, it can be seen that the total wear of cylindrical tool (portion IV) will result in a relative change in resonance frequency of the system equal to

$$\frac{\Delta f}{f} \simeq \tfrac{2}{3} n^{IV}$$

where n^{IV} is the wave number for portion IV. It is given that Δf is to be less than 1 kc/s. Thus

$$n^{IV} \geqslant \frac{3}{2} \frac{\Delta f}{f} = \frac{3}{2} \cdot \frac{1}{20} = 0.075$$

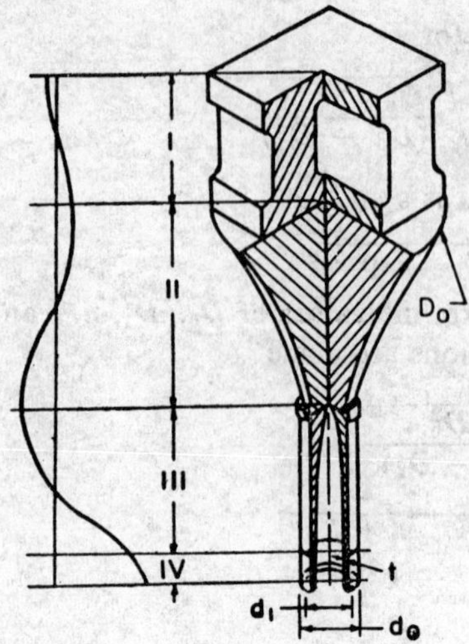

Fig. 2.7

or

$$l^{IV} \leqslant 0.075 \, \lambda^{IV} = 0.075 \left\{ \frac{C}{f} \right\}^{IV}$$

Assuming that the tool is made of brass

$$C = 3.42 \times 10^6 \text{ mm/s}$$

$$l^{IV} \leqslant 0.075 \, \frac{3.42 \times 10^6}{20 \times 10^3}$$

$$\leqslant 12.8 \text{ mm}$$

Take $l^{IV} = 12.5$ mm. If 1 mm is provided as the wall thickness of the tool, inner diameter d_1 would be $d_0 - 2t = 9-2 = 7$ mm.

$$\text{Length of the tool} = 12.5 \text{ mm}$$
$$\text{Outside diameter of the tool,} \quad d_0 = 9 \text{ mm}$$
$$\text{Inside diameter of the tool,} \quad d_1 = 7 \text{ mm}$$

Further, considering that portions II and III are made of steel $(C = 5.0 \times 10^6 \text{ mm/s})$, the design of portion III is carried out as follows

$$d_0 = 9 \text{ mm}$$
$$d_1 = 7 \text{ mm}$$
$$n^{III} = \tfrac{1}{2} - n^{IV} = \tfrac{1}{2} - 0.075 = 0.425$$

The amplification factor

$$K = \sqrt{\frac{d_0^2}{(d_0^2 - d_1^2)}} = 1.59099$$

$$l^{III} = n^{III} \cdot \frac{C}{f} \sqrt{\left(1 + \frac{\ln K}{2\pi n^{III}} \right)^2}$$

$$= 0.425 \times \frac{5 \times 10^6}{20 \times 10^3} \sqrt{1 + \left(\frac{0.464356}{2\pi \times 0.425} \right)^2}$$

$$= 107.84 \text{ mm}$$

Exponential factor

$$\beta = \frac{\ln K}{l^{III}} = \frac{0.464356}{107.84} = 4.3058 \times 10^{-3}\ \text{mm}^{-1}$$

The internal shape follows the law

$$d = d_0\ \sqrt{1 - \exp^{(-2\beta x)}}$$

$$= 9 \times \sqrt{1 - \exp^{-0.008611x}}$$

The dimensions of portion III are given in Table 2.3.

Table 2.3

x mm	$0.008611x$	$\sqrt{1-\exp^{-0.008611x}}$	d mm
0	0	0	0
10	0.08611	0.2872	2.585
20	0.17222	0.39775	3.58
30	0.25833	0.47714	4.29
40	0.3444	0.5397	4.858
50	0.4305	0.5915	5.323
60	0.5166	0.6352	5.72
70	0.6027	0.6728	6.055
80	0.6888	0.7055	6.35
90	0.7750	0.7344	6.61
100	0.8611	0.7598	6.83
107.8	0.9282	0.7776	7.00

Check

$$f_c = \frac{\beta C}{2\pi} = \frac{4.3058 \times 10^{-3} \times 5 \times 10^{6}}{2\pi} = 3.43 \times 10^{3}\text{c/s}$$

$$1.5 f_c = 5.15\ \text{kc/s}$$

$$f = 20\ \text{kc/s}$$

As $f > 1.5 f_c$, the concentrator will work satisfactorily.

The design of portion II, which is an ordinary circular half-wave concentrator, is as follows.

In order to completely utilize the radiating surface of the transducer, the diameter of portion II in contact with the transducer is made equal to the diagonal or the transducer section.

$$D_0 = 45 \sqrt{2} \simeq 63.6\ \text{mm}$$

$$n^{II} = \tfrac{1}{2}$$

Amplification factor $K = \dfrac{D_0}{d_0} = \dfrac{63.64}{9} = 7.07$

Length $\quad l^{II} = \dfrac{C}{2f} \sqrt{1 + \left\{ \dfrac{\ln K}{\pi} \right\}^2}$

$$= \frac{5 \times 10^6}{2 \times 20 \times 10^3} \sqrt{1 + \left\{ \frac{1.956}{\pi} \right\}^2} \simeq 147 \text{ mm}$$

Exponential factor $\beta = \dfrac{\ln K}{l^{II}} = \dfrac{1.956}{147} = 0.0133 \text{ mm}^{-1}$

The shape is given by

$$D = D_0 \, \exp^{(-\beta x)}$$
$$= 63.6 \, \exp^{(-0.0133x)}$$

Check

$$1.5 f_c = 1.5 \times \frac{\beta C}{2\pi} = \frac{1.5 \times 0.0133 \times 5 \times 10^6}{2\pi} = 15.87 \times 10^3 \text{ c/s}$$

$$f = 20 \text{ kc/s} > (1.5 f_c = 15.87 \text{ kc/s})$$

The calculations for the dimensions of portion II are given in Table 2.4.

Table 2.4

$\dfrac{x}{mm}$	$0.0133x$	$\exp^{(-0.0133x)}$	$\dfrac{D}{mm}$	$\dfrac{x}{mm}$	$0.0133x$	$\exp^{(-0.0133x)}$	$\dfrac{D}{mm}$
0	0	1	63.6	80	1.064	0.3450	21.95
10	0.133	0.8755	55.68	90	1.197	0.3020	19.21
20	0.266	0.7664	48.74	100	1.330	0.2644	16.82
30	0.399	0.6710	42.67	110	1.463	0.2315	14.72
40	0.532	0.5874	37.36	120	1.596	0.2027	12.89
50	0.665	0.5142	32.70	130	1.729	0.1775	11.28
60	0.798	0.4502	28.63	140	1.862	0.1553	9.88
70	0.931	0.3941	25.06	147	1.955	0.1415	9.00

Mechanics of Cutting

The mechanics of material removal from a surface by forcing sharp grains into it have been studied by many researchers [3, 4, 5, 6]. Based on certain assumptions, they have given estimates of the cutting rate. A brief discussion follows.

Theory of Miller [3]

The cutting rate for brittle materials depends on the size of chips and the rate at which these chips are formed. For ductile materials, it is assumed that the process involves the imbedding of abrasive particles in the surface by the plastic flow which leads to the work hardening of material at the surface. Material removal is affected thus by the process of chipping.

For deriving the expression for the rate of metal removal, Miller made the following assumptions:

(i) Cutting involves plastic flow of material.

(ii) Abrasive particles are cubes of side d.

(iii) All particles under the tool cut effectively.

It is considered that the volume rate of metal removal V can be expressed as a function.

$$V = v \text{ (PD) (TN) (WHR) (VC) (CR) (T)} \qquad 2.11$$

where

v — function

PD — plastic deformation for abrasive particles striking the work surface

TN — total number of abrasive particles striking per unit time

WHR — work hardening per unit of plastic deformation

VC — volume of work material chipped by each grain in each blow

CR — rate at which blows are struck

T — rate of flow of abrasive particles under the tool tip

The depth of penetration h of cubic particle of side d, under an applied impulsive stress σ can be taken as

$$h \propto \sigma \, d \frac{1}{\xi}$$

where ξ denotes the plastic flow stress for the work material, and is given by

$$\xi = \epsilon \, b \, G$$

where ϵ is a constant, b is Burger's vector, and G is the shear modulus for the material.

The volume of plastic deformation of grains due to penetration to depth h is given by

$$(PD) = K_1 h \, d^2 \qquad (2.12)$$

where K_1 is a constant. At the end of the cutting stroke, as the tool rises, an evacuated region is imagined to be formed under the tool and the slurry containing abrasive is supposed to begin to flow into this region. As it has only a fraction of a second, it is not able to fill the region. Considering a circular tool and using the equation of motion of a particle subjected to atmospheric pressure, the effective number of cutting particles N under the tool tip can be written as

$$N = \frac{\pi \, r \, a \, p'}{2f^2 \, \rho_a \, V' d^2(c+1)}(c) \qquad (2.13)$$

where

r — radius of the tool

a — amplitude of vibration

p' — atmospheric pressure

f — frequency of vibration

ρ_a — density of abrasive

V' — volume of slurry in the puddle

c — ratio (mass of abrasive/mass of carrier), that is, concentration

d — size of abrasive particles

The total number of abrasive strokes on the work piece per second is

$$(TN) = N \cdot f \tag{2.14}$$

The amount of work hardening per unit of plastic deformation can be taken to be inversely proportional to the work hardening capacity q and directly proportional to the speed of deformation.   Therefore

$$(WHR) = K_2 \cdot f/q \tag{2.15}$$

where K_2 is a constant.

The amount of metal removed by an abrasive grain in each blow is considered to be proportional to the grain volume, that is

$$(VC) = K_3 d^3 \tag{2.16}$$

where K_3 is a constant.

The rate at which blows are struck is the same as the frequency of vibration of the tool tip.   Thus

$$(CR) = f \tag{2.17}$$

Considering the flow of abrasive particles under the tool, as in Eq. 2.13, the rate at which the tool tip covers the abrasive particles can be written as

$$(T) = K_4 \ \frac{a \, p}{2 \, r \, d^3 \, f \, \rho_a} \ \frac{(c)}{V' \, (c + 1)} \tag{2.18}$$

where K_4 is a constant.

On substitution, the machining rate equation is obtained in the form

$$V = K_5 \frac{F \, d \, a \, p \, f}{q \, G \, b \, r \, \rho_a \, V'} \cdot \frac{c}{c+1} \tag{2.19}$$

where K_5 is an overall constant, and F is the average force on all particles under the tool, that is

$$F = N \, \sigma \, d^2$$

Comments on analysis by Miller   Equation 2.19 is to some extent in agreement with the experimental values of the ultrasonic machining rate. However, the assumptions on which the cutting rate expression has been derived have no logical justification.   For example, the assumption that the machining rate is dependent upon work hardening is very much doubtful. Miller considered the use of plastic materials only, whereas most materials worked by this process are brittle in nature, and so the rate cannot be dependent on plastic deformation.   The estimation of the number of particles under the tool is also based on many hypothetical considerations.   The assumptions that all particles of abrasive are of equal size and are cubes, and that under the tool tip they all take part in cutting, are not realistic. The volume of material cut is dependent on amplitude as well as force, but in Miller's analysis, amplitude affects the rate only through the number of particles under the tool.

Theory of Shaw [4]

Shaw considered the process of cutting by ultrasonics to take place by four possible actions:

   (i) The abrasive grains are thrown onto the work surface and then the surface is chipped.

  (ii) The abrasive grains are hammered onto the work surface.

 (iii) Erosion or caviation occurs under the tool and causes material removal.

 (iv) Chemical corrosion associated with abrasive carrying media causes material removal.

Actions (iii) and (iv) are not of much importance, because when abrasive is not present, the machining rates are observed to be extremely low.

Shaw considered all the abrasive particles to be spheres of diameter d. These particles hit the work surface and generate a crater or identation of height h. If the depth of penetration is sufficient to cause rupture, it can be assumed that the particle chipped out has volume approximately proportional to the diameter of identation. The amount of material removed per active grit can then be written as

$$V_g = K_1 (h \cdot d)^{3/2} \tag{2.20}$$

where K_1 is a constant.

In one cycle of tool, the number of impacts depends on the number of grits that come underneath the tool at any instant, and can be written as an inverse function of the cross sectional area of the grits. Thus, the number of impacts per cycle

$$N = K_2 \frac{1}{d^2} \tag{2.21}$$

where K_2 is a constant.

Taking K as the probability of a grit under the tool being active, the volume of material removed per cycle then is

$$V_c = \frac{K \, K_2 \, K_1 \, (h \cdot d)^{3/2}}{d^2} = K \, K_1 \, K_2 \sqrt{\frac{h^3}{d}} \tag{2.22}$$

and the volume rate of material removal per unit time is

$$V = K \, K_1 \, K_2 \sqrt{\frac{h^3}{d}} \cdot f \tag{2.23}$$

For calculating the cutting rate by Eq. 2.23, h, the depth of penetration of an abrasive particle is unknown. To determine this, Shaw made two models.

Model 1 Consider the grains to be thrown onto the work surface. The displacement y of the tool from mean position ($t = 0$) at time t can be written as

$$y = \frac{a}{2} \sin (2\pi f t) \tag{2.24}$$

where a and f are respectively the amplitude and the frequency of the tool vibration.

The velocity of tool at any time t is

$$\dot{y} = \pi\, a\, f \cos\left(2\pi f t\right) \tag{2.25}$$

And the maximum velocity

$$\dot{y}_{\text{max}} = \pi\, a\, f \tag{2.26}$$

This is the velocity at which the grits leave the tool. The kinetic energy of a grit with ρ_a as its density is

$$\text{K. E.} = \tfrac{1}{2}\left\{\frac{\pi}{6}\, d^3\, \rho_a\right\} \pi^2\, a^2\, f^2 \tag{2.27}$$

This energy is absorbed by the work surface to bring a particle to rest after penetrating a distance h. If F denotes the force acting on a grain as it penetrates the work surface, then

$$\text{work done} = \tfrac{1}{2}\, F\, h \tag{2.28}$$

and mean stress acting on work surface

$$\sigma = \frac{F}{A} = \frac{F}{\pi h\, d} \tag{2.29}$$

where A is the projected area of a grit. Substituting the value of F from Eq. 2.28 and equating with Eq. 2.27

$$h = \pi\, a f d \sqrt{\frac{\rho_a}{6\sigma}} \tag{2.30}$$

Model 2 Assume that the grains are hammered into the work piece surface. Force F applied to the abrasive grain acts for only a short time ΔT in one complete cycle of duration T. The mean static force F can be written as

$$F = \frac{1}{T}\int_{0}^{T} F(t)\cdot dt \tag{2.31}$$

where $F(t)$ is the force at any instant t. The momentum equation may be written approximately as

$$\int_{0}^{T} F(t)\, dt \simeq \frac{F}{2}\cdot \Delta T \tag{2.32}$$

where F is the force of the tool during the penetration of the grains.

When hammered, the grains, being in contact with both the tool and the work surfaces, will penetrate both the surfaces. If h and h' represent the depths of penetration into the work piece and tool surface respectively, the total movement of tool h while hammering is $h = (h+h')$ and thus

$$\Delta T \simeq \frac{h}{a}\cdot \frac{T}{2} \tag{2.33}$$

which gives to an approximation

$$F = F \frac{4a}{h} \qquad (2.34)$$

If N is the number of grains under the tool, the stress on the work surface and the tool can be written as

$$\sigma = \frac{F}{N(\pi h d)} \qquad (2.35)$$

$$\sigma' = \frac{F}{N(\pi h' d)} = \sigma \frac{h}{h'} \qquad (2.36)$$

Substituting F from Eq. 2.34 and N from Eq. 2.21 in Eq. 2.35

$$\sigma = \frac{4 F a d}{\pi h^2 K_2 (h'/h + 1)} \qquad (2.37)$$

From Eq. 2.36, $h'/h = \dfrac{\sigma}{\sigma'} = j$, that is, j can be taken as the ratio of the hardness of work material to that of the tool material. Equation 2.37 gives

$$h = \sqrt{\frac{4 F a d}{\pi \sigma K_2 (j + 1)}} \qquad (2.38)$$

On the basis of the two mechanisms considered, the rate of cutting can thus be estimated by substituting Eq. 2.30 and Eq. 2.38 respectively in Eq. 2.23.

(i) When abrasive particles are considered to be thrown at the work piece

$$V_1 = K K_1 K_2 \left[\frac{\pi^2 a^2 \phi_a}{6 \sigma} \right]^{3/4} d f^{5/2} \qquad (2.39)$$

(ii) When abrasive particles are considered to be hammered into the work piece surface

$$V_2 = K K_1 K_2 \left[\frac{4 a F}{\sigma \pi K_2 (j + 1)} \right]^{3/4} d^{1/4} \cdot f \qquad (2.40)$$

By putting numerical values in these equations it can be seen that the cutting rate determined from Eq. 2.39, that is, by the thrown-abrasive mechanism, is much smaller than that from Eq. 2.40, that is, by the hammering mechanism. Therefore, the abrasive throwing mechanism can be neglected while estimating the cutting rate. Further, the depth of penetration of abrasive grains is actually very small compared to their diameter. It can thus be assumed that the cutting action may also be a result of irregularity or roughness on the surface of grits.

Theory of Kazantsev et al. [5, 6]

The analysis proposed by Kazantsev et al. assumes that material is removed by brittle fracture. Two important features of the process are considered here, namely, the grits are of irregular shape and there is a distribution of particle size. The model on which this anlaysis is based is shown in Fig. 2.8. When the tool comes down to its lowest point, it causes some grains to penetrate both the tool and the work surface.

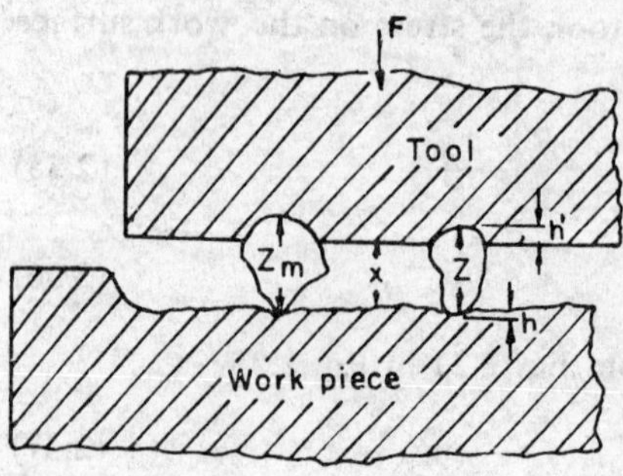

Fig. 2.8 A model for analysis of material removal rate in ultrasonic machining, as proposed by Kazantsev, et al.

Let there be N number of particles between the tool and workpiece. Let Z_m denote the maximum size of particles and the function ψ_z describe the size distribution of particles. Assuming the abrasive particles to be incompressible, the total penetration into the tool and work piece of any grit size Z is $(Z-x)$. If the force on any grit is assumed to be a function of the penetration, that is, force on a grit is $\phi_z (Z-x)$, the total force acting on all particles

$$F = N \int_{x}^{Z_m} \phi_z (Z-x) \cdot \psi_z (Z) \, d Z \tag{2.41}$$

On similar lines, the volume of material removed per cycle can be expressed as a function of Z. Assuming that the volume of material taken away by an abrasive particle is a function of the penetration $(Z-x)$, namely $\theta_z (Z-x)$, the total volume of material machined in one cycle is

$$V = N \int_{x}^{Z_m} \theta_z (Z-x) \cdot \psi_z (Z) \cdot d Z \tag{2.42}$$

Before Eqs. 2.41 and 2.42 can be used, functions $\psi_z (Z-x)$ and $\theta_z (Z-x)$ have to be determined.

Kazantsev and Rosenberg [5] made a statistical analysis of a number of abrasive powders and empirically determined the size distribution of grains as

$$\psi_z (Z) = \frac{1.095}{Z} \left\{ 1 - \left[\frac{Z - Z}{Z} \right]^2 \right\}^3 \tag{2.43}$$

On the basis of the work of Shreiner [7] and Koifman [8], they estimated the functions ϕ_z and θ_z. The final cutting rate equation was obtained after a tedius numerical analysis. It contained the terms for the hardness of tool, work and abrasive materials and a number of constants. They verified tlc equation with their experimental test results and reported that good correlation existed between the two.   Their analysis is, however, very cumbersome to use.

Effect of Parameters

Effect of Amplitude and Frequency of Vibrations
Miller [3] found that for a given material, cutting rate R bears a linear relationship with amplitude.   His data, however, had much scatter (Fig. 2.9). Goetze [9] also claimed the same.   According to him, cutting rate increases linearly with an increase in both amplitude, a and frequency f and that after a certain critical abrasive/water ratio, the term $(R/a\,f)$ is almost constant.

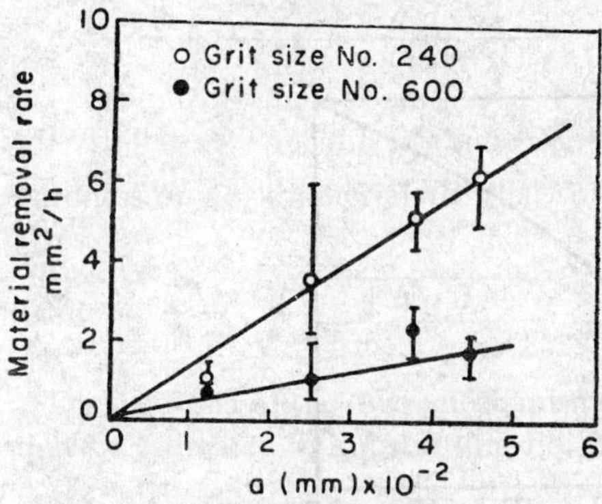

Fig. 2.9 Relation between amplitude of vibrations and material removal rate. Abrasive: Boron carbide; Frequency: 25 kc/s; Tool material: Ketos; Work material: Steel

Rosenberg et. al. [5] found the material removal rate R to be proportional to the square of amplitude; Pentland and Ektermanis [10] in their results, confirmed the square law. Shaw's [4] results, however, indicated the variation of removal rate to be proportional to (amplitude)$^{3/4}$. The results of Neppiras and Foskett [11] indicated non-linear relationships between R and frequency as well as amplitude (Fig. 2.10). Increasing the amplitude tends to increase the surface roughness, but the effect is minimal.

There are discrepancies in the reported results of the effects of frequency on the material removal rate.   Shaw and Goetze predicted a linear relationship (Fig. 2.11).   But Neppiras and Foskett indicated a non-linear relationship (Fig. 2.10).   Rosenberg [5] observed frequency dependence to be interrelatad with work and tool failure characteristics. He found a higher rate of increase in the removal rate with frequency in brittle materials than in ductile materials (Fig. 2.12).   Markov [12] has shown (Fig. 2.13) that material removal rate is directly proportional to the particle velocity, that is, amplitude/resonant frequency bears a linear relationship to machining rate.   It thus implies that the frequency used for the machining process must be the resonant frequency of the acoustic system in order to obtain greatest amplitude at the tool tip and thus achieve the maximum utilization of the acoustic system.

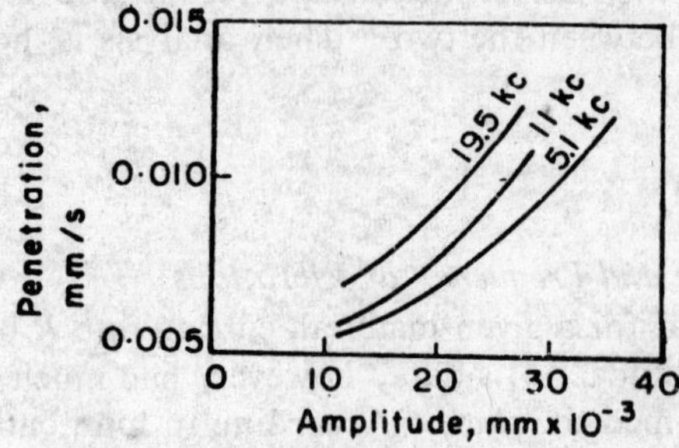

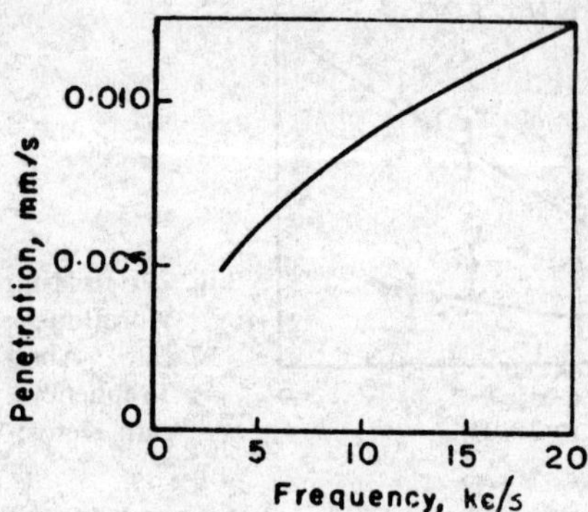

Fig. 2.10 Results of experiments conducted
by Neppiras and Foskett [11]

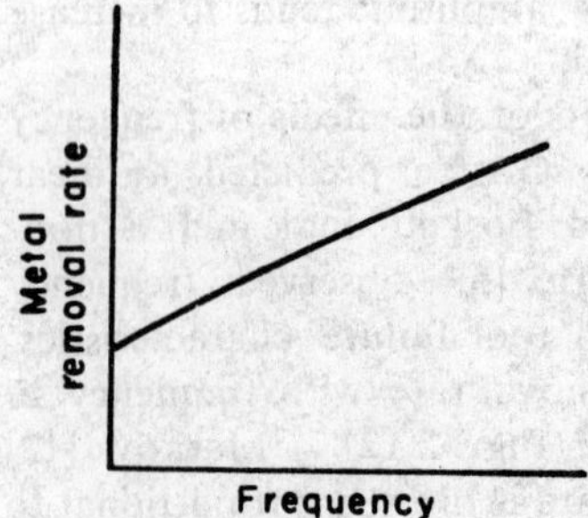

Fig. 2.11 Results of experiments
conducted by Shaw

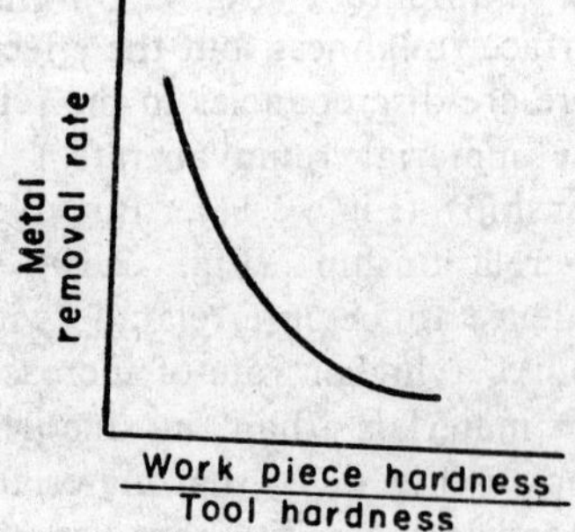

Fig. 2.12 Relationship between
material removal rate
and hardness ratio of
work piece and tool
materials

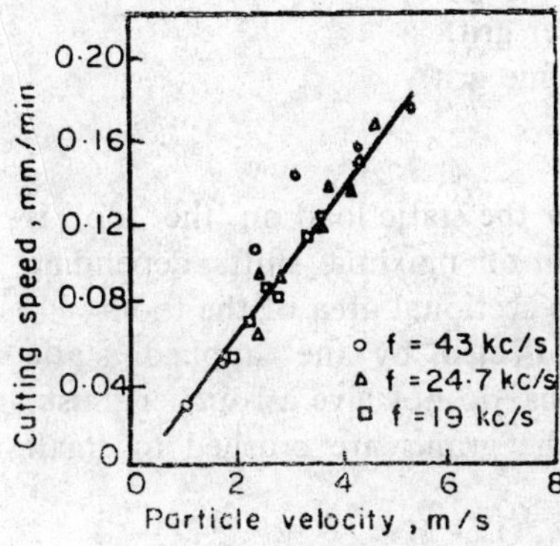

Fig. 2.13 Relation between abrasive grain velocity and the cutting speed

Effect of Grain Diameter

Goetze claimed that cutting rate increases linearly with grain size, but the findings of Neppiras and Foskett [11] indicated a non-linear effect of grain diameter on the removal rate. There is a limit to the effect of grain size on the rate as a very coarse powder may even cause a fall in rate (Fig. 2.14). The optimum size is, however, governed by the amplitude of tool vibration. As the grit size becomes comparable with amplitude,

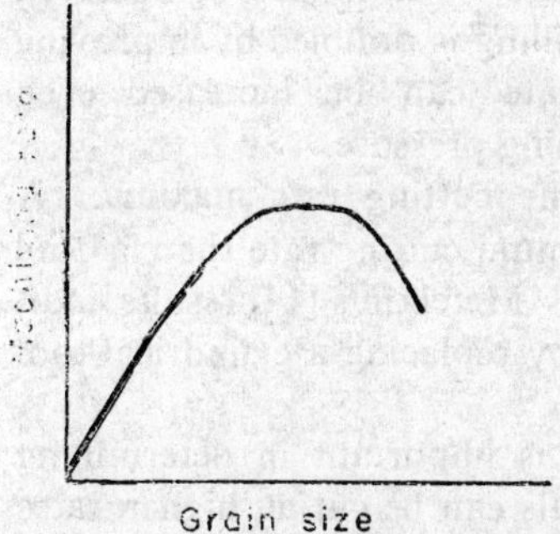

Fig. 2.14 Relation between abrasive grain size and the material removal rate

the optimum condition is achieved. Surface finish is greatly influenced by the grain size. The data obtained by Neppiras and Foskett on glass and tungsten carbide work materials has indicated that the bottom of the hole is smoother than the sides. A reason for this may be a cvitation streamers drawing abrasive grains down into the cutting zone which mark the surface of the sides.

Grain size determines the accuracy of the cavity configuration in ultrasonic machining. The hole cut is larger than the tool owing to the flow of abrasive along the sides of the hole to the bottom face of the tool. For accuracy and good surface fiinish, it is better to use a set of tools and more than one size of abrasive grits in stages, as follows:

Stage 1   undersize tool, high frequency, coarse grit.
Stage 2   undersize tool, high frequency, finer grit.
Stage 3   full size tool, low frequency, very fine grit.

Effect of Applied Static Load

The machining rate reaches a maximum as the static load on the tool is increased. As shown in Fig. 2.15, the point of maxima shifts, depending on the amplitude of vibration and the cross sectional area of the tool.

Surface finish is found to be little affected by the applied static load. Higher loads, contrary to expectations, do not give a rougher finish. Surface finish, in fact, improves because the grains are crushed to small size with higher loads.

Effect of Slurry, Tool and Work Material

It has been found that a rise in cutting rate can be achieved with an increase in slurry concentration. Neppiras and Foskett have given details of the influence of material and abrasive properties on the material removal rate. Neppiras further established that saturation occurs when the volume of the slurry is 30 to 40 per cent abrasive/water mixture (Fig. 2.16). Experimental results have shown a sharp drop in the material removal rate with increasing viscosity.

The pressure with which the slurry is fed into the cutting zone has a remarkable effect on the material removal rate. Pentland [10] observed that the material removal rate in ultrasonic drilling is doubled by improving slurry circulation. The material removal rate can be increased even ten times by supplying the slurry at an increasing pressure.

The shape of the tool face also affects the cutting rate maxima. A narrow rectangular tool gives a greater maximum cutting rate than a tool of the same area with a square cross section. Metelkin's [14] results indicated a rise in the cutting rate by 50 per cent by replacing a cylindrical tool with a conical one.

The brittle behaviour of the work material is important in determining the cutting rate. Brittle non-metallic materials can be cut at higher rates than ductile materials. Figure 2.16 gives the qualitative relationship between the material removal rate and work piece/tool hardness.

Effect on Work Material

Since the cutting forces involved in ultrasonic machining are not large, no mechanical stresses are setup which could cause warping or other residual deformation. The microstructure of work material is not affected as the temperature generated at the cutting area is not appreciable. But if the flow of abrasive slurry is impaired, high heat may develop, which may sometimes lead to cracking.

In drilling through holes, chipping may occur at the exit side. A simple way to overcome this is to fasten the work piece to a base (usually of glass)

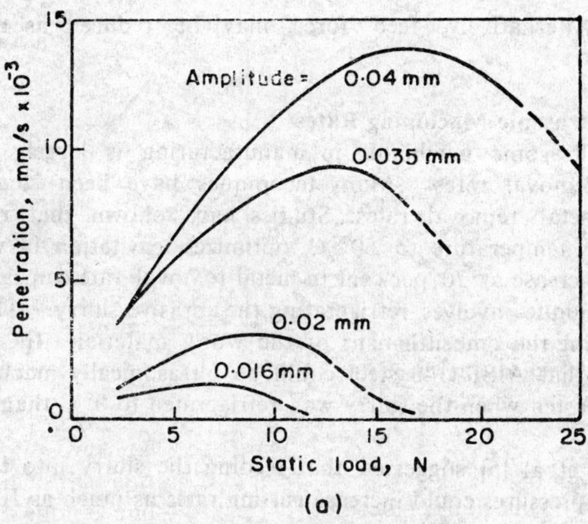

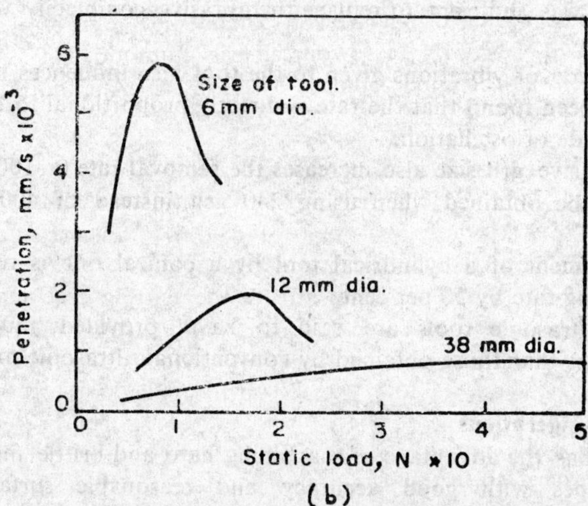

Fig. 2.15 Effect of applied static load on the material removal rate at different amplitudes of vibration and for different sizes of tool

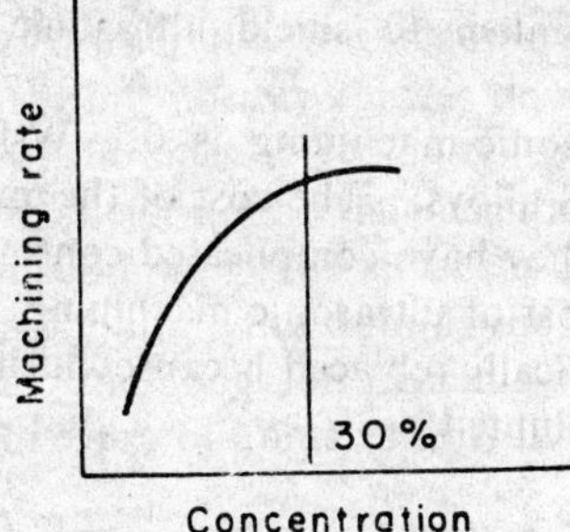

Fig. 2.16 Effect of slurry concentration on material removal rate

with wax. Alternatively, feed force may be reduced as the exit is approached.

Increasing Ultrasonic Machining Rates

Interest in ultrasonic machining in manufacturing is largely stifled by poor metal removal rates. Many techniques have been suggested for improving metal removal rates. Studies have shown that raising the abrasive slurry temperature to 50° C optimizes cavitation in water, and as high an increase as 70 per cent in metal removal rate can be achieved. Another technique involves refrigerating the abrasive slurry. This has the effect of causing the embrittlement of the work material. In one study it was found that AISI 4140 steel could be ultrasonically machined 20 to 30 per cent faster when the slurry was refrigerated to 0° C than at room temperature.

Rosenberg et al [5] suggested that feeding the slurry into the cutting zone at high pressures could increase cutting rates as much as 10 times the normal ultrasonic rates. The maximum possible rate could be achieved if the flow rate was sufficient to replace the abrasive completely within one cycle.

The amplitude of vibrations given to the tool also influences the cutting rate. It has been found that the rate is directly proportional to the square of the amplitude of oscillation.

Larger abrasive grit size also increases the removal rate (a 100 per cent increase may be obtained when using 240 grit instead of 400 grit size abrasive).

The replacement of a cylindrical tool by a conical one is reported to raise the cutting rate by 50 per cent.

Rotating ultrasonic tools are said to have provided much higher machining rates than those obtained by conventional ultrasonic machining.

Economic Considerations

The process has the advantage of machining hard and brittle materials to complex shapes with good accuracy and reasonable surface finish. Considerable economy results from the ultrasonic machining of hard alloy press tools, dies and wire drawing equipment on account of the high wear resistance of tools made of these alloys.

The machines have no high speed moving parts. Working on machines is not hazardous, provided care is taken to shield ultrasonic radiations from falling on the body.

The power consumption of ultrasonic machining is 0.1 W-h/mm^3 for glass and about 5 W-h/mm^3 for hard alloys. The cost of the manufacture and use of the tools, particularly if they have complicated contours, is very high. Another item adding to the cost of ultrasonic machining is abrasive. The abrasive slurry has to be periodically replaced because during use the particles are eventually broken and blunted.

Ultrasonic machines are not yet completely reliable; failure sometimes occurs on account of faults in acoustic head, pump or generator.

It is probable that with more research in the near future on techniques and machines, the process will have more economic advantages.

Applications of Ultrasonic Machining

When compared with other modern machining techniques described in this book, this method of machining is not limited by the electrical or chemical characteristics of work materials, which makes it suitable for application on to both non-conductive and conductive materials. Tungsten and other hard carbides and gem stones, such as synthetic ruby (for the preparation of jewels for watch and timer movements) are being successfully machined by this method.

The process is particularly suited to make holes with a curved axis of any shape that can be made on the tool. The range of shapes can be increased by moving the work piece during cutting. Figure 2.17 shows various operations that can be performed with ultrasonic machining.

The smallest hole that can currently be cut by the ultrasonic machining method is 0.050 mm in diameter, the hole size being limited by the strength of the tool and the clearance required for the flow of abrasive. The largest diameter solid tool reported to have been employed in some applications is 100 mm in diameter.

Limitations of the Process

The major limitation of the process is its comparatively low metal cutting rates. The depth of the cylindrical holes is presently limited to 2.5 times

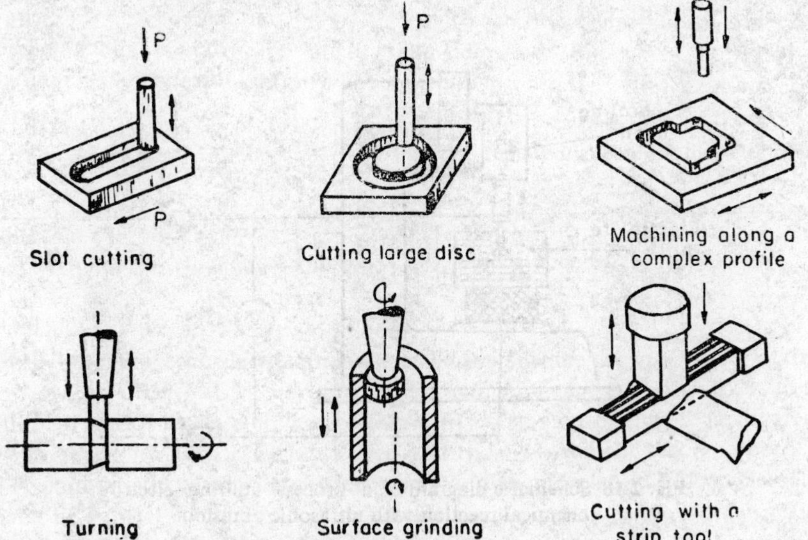

Fig. 2.17 Some applications of ultrasonic machining

the diameter of the tool. Tool wear increases the angle of the hole, while sharp corners become rounded. This implies that tool replacement is essential for producing accurate blind holes.

Due to the problem of fewer active grits coming under the tool's centre on account of ineffective slurry distribution, the bottom of a cavity cannot usually be machined flat. Sometimes the accuracy of a machined surface is lost due to the presence of strong lateral vibrations which are set up if the shape of the tool cross section is such that the centre of gravity is not on the centre line. In such a case, the only solution is to redesign the tool.

The tendency for holes to 'break out' at the bottom owing to static load and amplitude is another limitation. This can probably be overcome by programming the feed force and amplitude of tool vibration.

Recent Developments

Mullard Research Laboratories, USA, have developed a process that combines electrochemical reaction with ultrasonic abrasion (Fig. 2.18). Using a 60 W ultrasonic drill and abrasive suspended in an alkaline electrolyte, Mullard researchers have reported that tool steel can be machined nine times faster than by ultrasonics alone.

Engis Limited of England has developed the Disonic Die Ripper in which the diamond-plated tool oscillates at ultrasonic speed as well as rotates at high speed (5000 rpm) in a liquid to rapidly remove material from tungsten carbide dies. The use of diamond-plated tool eliminates the need of abrasive slurry and frequent re-grinding of steel tools. The oscillating-cum-rotational system is claimed to increase the material removal rate several times.

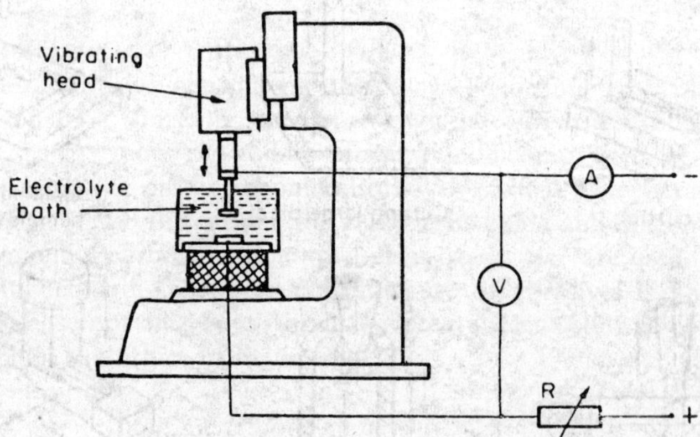

Fig. 2.18 Schematic diagram of a process utilizing electro-chemical reaction with ultrasonic abrasion

ABRASIVE JET MACHINING

Abrasive jet machining (AJM) uses a stream of fine grained abrasive mixed with air or some other carrier gas at high pressure. This stream is directed by means of a suitably designed nozzle on to the work surface to be machined. Metal removal occurs due to erosion caused by the abrasive particles impacting the work surface at high speed. Figure 2.19 shows a schematic diagram of the working of this process.

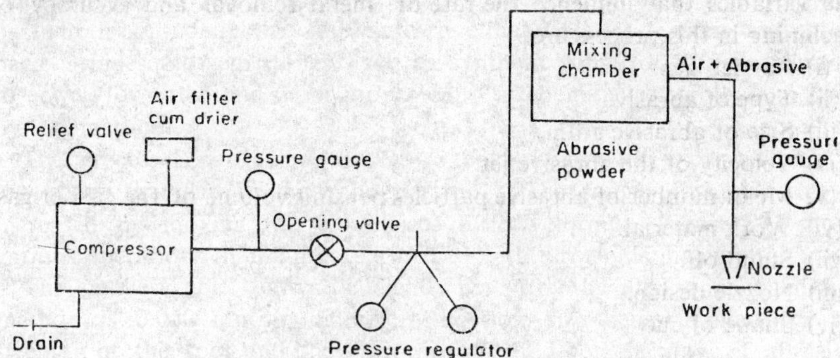

Fig. 2.19 Schematic diagram of abrasive jet machining

Applications

Lavoie [18] and Ingulii [19] have both mentioned a number of interesting applications of AJM. This process has been successfully applied for the following operations:

  (i) Removing flash and parting lines from injection moulded parts.

  (ii) Deburring and polishing plastic, nylon and teflon components.

  (iii) Cleaning metallic mould cavities which otherwise may be inaccessible.

  (iv) Cutting thin sectioned fragile components made of glass, refractories, ceramics, mica, etc.

  (v) Producing high quality surface.

  (vi) Removing glue and paint from paintings and leather objects.

  (vii) Reproducing designs on a glass surface with the help of masks made of rubber, copper, etc.

  (viii) Frosting interior surfaces of glass tubes.

  (ix) Etching markings on glass cylinders.

Advantages and Disadvantages

The advantage of this process lies in its ability to machine brittle materials with thin sections, especially in areas which are inaccessible by ordinary methods. Absence of tool work contact and metal removal at microscopic scale leads to very little or no heat generation, resulting in insignificant surface damage. The process is characterized by low capital investment and low power consumption. However, the application of AJM is restricted to brittle materials because of the lower rates of metal removal attainable in cases of ductile materials. Sometimes, parts machined by this process have to undergo an additional operation of cleaning as there is a possibility of the abrasive grains sticking to the surface. The machining accuracy is poor and the nozzle wear rate is high. Further, the process tends to pollute the environment.

Variables in AJM

The variables that influence the rate of metal removal and accuracy of machining in this process are:

   (i) Carrier gas.

  (ii) Type of abrasive.

 (iii) Size of abrasive grain.

 (iv) Velocity of the abrasive jet.

  (v) Mean number of abrasive particles per unit volume of the carrier gas.

 (vi) Work material.

(vii) Stand off.

(viii) Nozzle design.

 (iv) Shape of cut.

Carrier Gas

Carrier gas, to be used in abrasive jet machining, must not flare excessively when discharged from the nozzle into the atmosphere. Further, the gas should be nontoxic, cheap, easily available and capable of being dried and cleaned without difficulty. The gases that can be used are air, carbon dioxide or nitrogen. Air is most widely used owing to easy availability and little cost. All abrasive powders supplied by the manufacturers can be run with clean shop air, provided air filters have been installed in the air lines.

Types of Abrasives

The choice of abrasive depends on the type of machining operation, for example, roughing, finishing, etc., work material and cost. The abrasive should have a sharp and irregular shape and be fine enough to remain suspended in the carrier gas and should also have excellent flow characteristics.

The abrasives used for cutting are aluminium oxide and silicon carbide, whereas sodium bicarbonate, dolomite, glass beads, etc. are used for cleaning, etching, deburring and polishing. Re-use of abrasives is not recommended because not only does its cutting ability decrease, but contamination also clogs the orifice of the nozzle.

Grain Size

The rate of metal removal depends on the size of the abrasive grain. Finer grains are less irregular in shape, and hence, possess lesser cutting ability. Moreover, finer grains tend to stick together and choke the nozzle. The most favourable grain sizes range from 10 to 50 μ (Table 2.5). Coarse

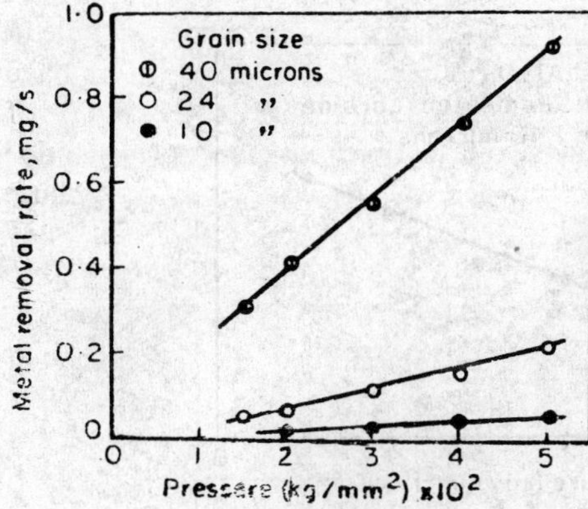

Fig. 2.20 Effect of abrasive jet pressure on the material removal rate. Abrasive: Aluminium oxide; Work material: Glass

grains are recommended for cutting, whereas finer grains are useful in polishing, deburring, etc. The effect of grain size on the rate of metal removal vis-a-vis abrasive jet pressure is shown in Fig. 2.20.

Table 2.5

Abrasive	Grain size	Applications
Aluminium oxide (Al₂O₃)	12, 20, 25 μ	Cutting and grooving
Silicon carbide (SiC)	25, 40 μ	-do-
Sodium bicarbonate	27 μ	Light finishing below 50° C
Dolomite	200 mesh (Approx.)	Etching and polishing
Glass beads	0.635 to 1.27 mm	Light polishing and fine deburring

Jet Velocity

The kinetic energy of the abrasive jet is utilized for metal removal by erosion. Finnie and Sheldon [20] have shown that for erosion to occur, the jet must impinge the work surface with a certain minimum velocity. For the erosion of glass by silicon carbide (grain size 25 μ), the minimum jet velocity has been found to be around 150 m/s.

The jet velocity is a function of the nozzle pressure, nozzle design, abrasive grain size and the mean number of abrasives per unit volume of the carrier gas. Figures 2.20 and 2.21 show the effect of nozzle pressure on the rate of metal removal.

Mean Number of Abrasive Grains per Unit Volume of the Carrier Gas

An idea about the mean number of abrasive grains per unit volume of the carrier gas can be obtained from the mixing ratio M. It is defined as

$$M = \frac{\text{Volume flow rate of the abrasive per unit time}}{\text{Volume flow rate of the carrier gas per unit time}}$$

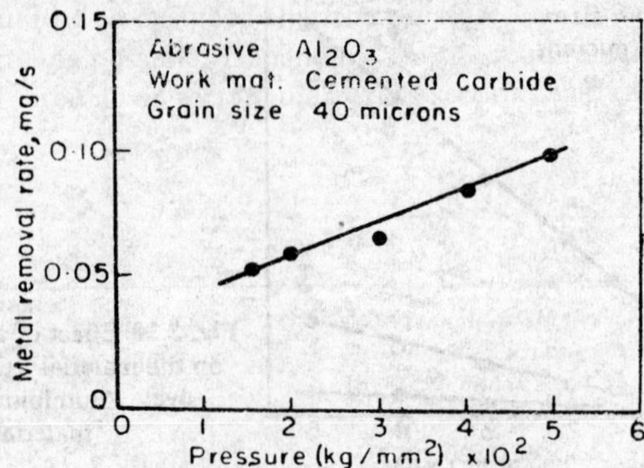

Fig. 2.21 Effect of abrasive jet pressure on the
material   removal   rate.   Abrasive:
Aluminium oxide;   Workmaterial:
Cemented carbide

A large value of M should result in higher rates of metal removal but a large abrasive flow rate has been found to adversely influence jet velocity, and may sometimes even clog the nozzle. Thus, for the given conditions, there is an optimum mixing ratio that leads to a maximum metal removal rate.

Work Material
AJM is recommended for the processing of brittle materials, such as glass, ceramics, refractories, etc.   Most of the ductile materials are practically un-machinable by AJM. The rate of metal removal has been found to depend upon the Mohr's hardness of the material to be machined (Figs. 2.20 and 2.21).

Stand off Distance (SOD)
Stand off is defined as the distance between the face of the nozzle and the working surface of the work. SOD has been found to have considerable effect on the rate of metal removal as well as the accuracy. A large SOD results in the flaring up of the jet which leads to poor accuracy. Figure 2.22 shows the effect of the SOD on the machining accuracy.

  Figure 2.23 shows the relationship between the SOD and the rate of material removal. Small metal removal rates at a low SOD is due to a reduction in nozzle pressure with decreasing distance, whereas a drop in material removal rate at large SOD is due to a reduction in the jet velocity with increasing distance.

Nozzle Design
The nozzle has to withstand the erosive action of abrasive particles, and hence, must be made of materials that can provide high resistance to wear.

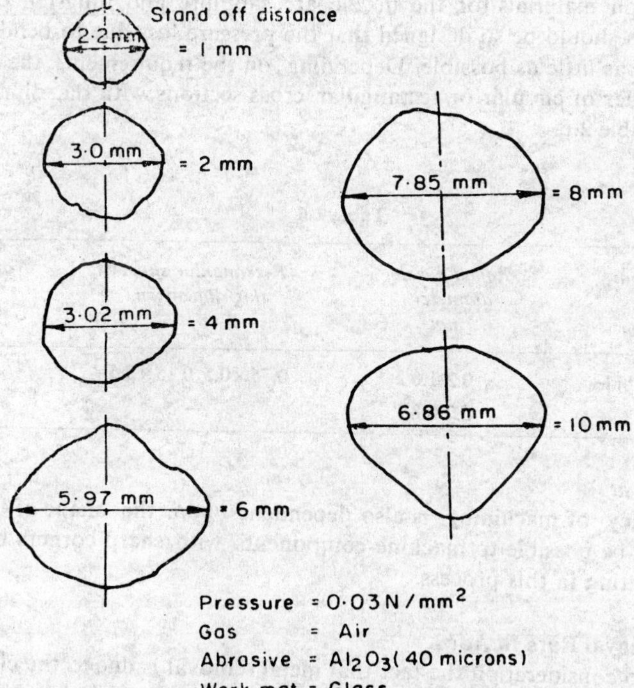

Fig. 2.22 Effect of stand off distance on the accuracy of shape produced

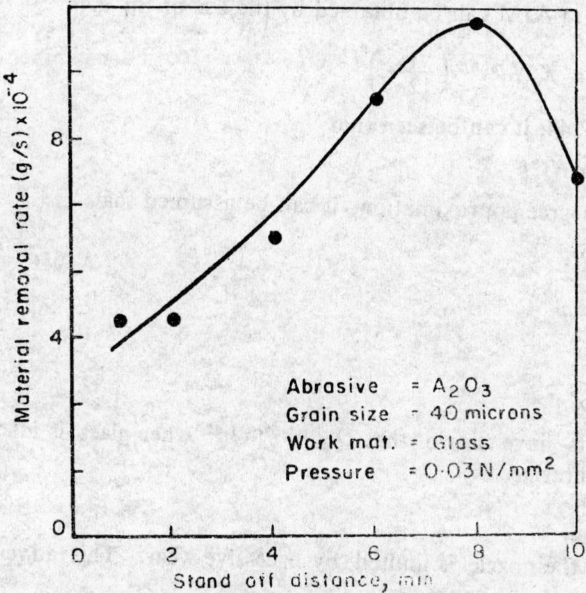

Fig. 2.23 Effect of stand off distance on material removal rate

The common materials for the nozzle are sapphire and tungsten carbide. The nozzles should be so designed that the pressure loss due to bends, friction, etc. is as little as possible. Depending on the requirements, the nozzles may be either of circular or rectangular cross sections with the dimensions given in Table 2.6.

Table 2.6

Nozzle material	Round nozzle diameter mm	Rectangular shape; slot dimension mm	Nozzle life h
Tungsten carbide	0.2–1.0	$0.75 \times 0.5 - 0.15 \times 2.5$	12–30
Sapphire	0.2–0.8	—	300

Shape of Cut

The accuracy of machining is also dependent upon the shape of the cut. It may not be possible to machine components with sharp corners because of stray cutting in this process.

Metal Removal Rate in AJM

Taking into consideration the fact that metal removal is due to the chipping of the work surface brought about by the impacting abrasive particles. Sarkar and Pandey [22] proposed a mathematical model for the computation of the metal removal rate in AJM. They have shown that the rate of metal removal Q in AJM can be obtained by the use of Eq. 2.44.

$$Q = K_4 N d^3 v^{3/2} \left(\frac{\rho_a}{12\sigma_y} \right)^{3/4} \tag{2.44}$$

From Eq. 2.44, it can be seen that

$$Q \propto N v^{3/2} \tag{2.45}$$

As a first degree approximation, it can be assumed that

$$N \propto p \tag{2.46a}$$

$$v \propto p^{\frac{1}{2}} \tag{2.46b}$$

Therefore

$$Q \propto p^{1 \cdot 75} \tag{2.47}$$

Experiments have shown that $Q \propto p^{1 \cdot 07 \sim 1 \cdot 42}$ when glass is machined with Al_2O_3 as abrasive.

Nozzle Wear

The life of the nozzle is limited by excessive wear. This adversely affects the accuracy of working and the metal removal rate.   Figure 2.24 shows that nozzle wear in the case of AJM is not uniform.

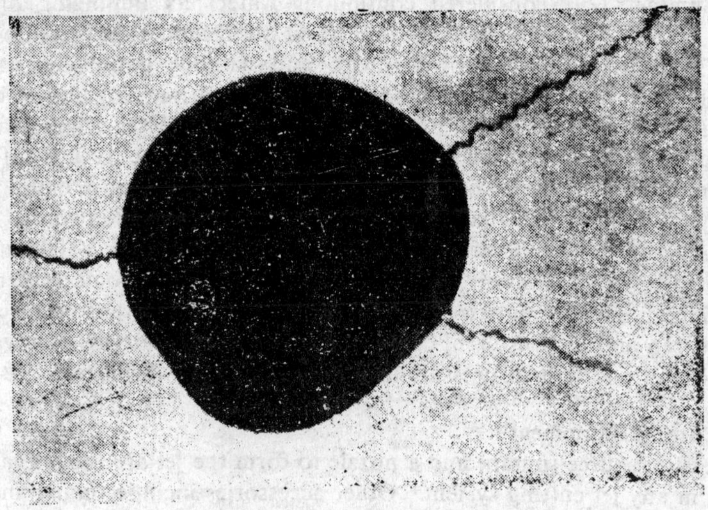

Fig. 2.24   Shape of worn out carbon steel nozzle after working for 30 minutes. Abrasive: Aluminium oxide

WATER JET MACHINING

Using a jet of water to cut a sheet of metal may sound impossible, but it is actually based on a principle we learn early in our lives. A simple example of this is when a finger is put over part of a tap, the stream of water flows with higher pressure so that it washes away mud far more effectively, giving a jet-cleaned item.

In scientific terms, if the jet of water is directed at a target in such a way that, on striking the surface, the high velocity flow is virtually stopped, then most of the kinetic energy of the water is converted into pressure energy. In fact, in the first few milliseconds after the initial impact of the jet on the target, before the lateral flow of water is initiated, a transient pressure as much as three times the normal stagnation pressure, may be generated. Erosion occurs if the local fluid pressures exceed the strength of the bond binding together the materials making up the target. In other words, liquid jet cutting removes material primarily by the mechanical action of a high velocity stream impinging on a small area. whereby its pressure exceeds the flow pressure of the material being cut.

Over the past three decades, due to the potential for reducing dust and associated hazards while cutting coal or rock, a number of studies using liquid jets have been made on these materials. Farmer and Attewell [28] reported the results of water jet impinging on sandstone.

The system used was a pulsed jet with velocity up to 500 m/s, and the effects of velocity on penetration were reported. A study by Brook and Summers [29] considered continuous water jets impinging on sandstone targets. The effects of varying stand off distances at pressures up to 92 MN/m² were reported for jets with and without a polymer additive. Pulsed water jets have been used in rock excavation [30] and machining aluminium and lead [31]. Franz [32] has reported the importance of improved coherence of the liquid jet and has given the results of using the liquid jet with the addition of a polymer. The use of a liquid jet for cutting materials other than coal or rock has also been studied by several researchers [32, 33]. Cutting capability at pressures up to 10,000 atm was reported for a wide variety of target materials, including wood, lead, rubber, aluminium, copper and steel [34]. A more recent study has been reported by Neusen and LaBrush [35] on the effectiveness of material removal as a function of nozzle inlet pressure and nozzle-to-target distance.

Jet Cutting Equipment

A pressure source (pump) and a nozzle to form the jet are the fundamental items in any jet cutting system. Other accessories include tubing and their fittings and valves. These are discussed briefly.

(i) *Pump* Pressurizing a liquid to the 1500-4000-MN/m² range is usually accomplished either by direct mechanical drive applied to a small diameter plunger or by an intensifier where an intermediate pressure fluid drives a large area piston, which in turn, drives a small diameter ram which pumps the cutting fluid. At these pressures, the main problem is sealing the high pressure rams, and at higher pressures, fatigue failure of the mechanical components can limit the life of the equipment. Some solutions have been found for sealing problems. One solution is to allow high pressure seals to be changed quickly by making them easily available; this means that the intensifier is out of commission for the least possible time. Alternatively, the conventional fabric-backed, synthetic rubber seal can be lubricated by adding soluble oil up to 5 per cent to the water being pumped. However, this oil may be incompatible with the material to be cut and the disposal of the waste fluid could also cause considerable difficulties.

The reciprocating ram can be surrounded by a long, close-fitting sleeve. With correct design and precision manufacture, the fluid leakage through this clearance seal can be as low as 2 per cent of the rated delivery of the unit at pressures of 30 kN/cm³. Made from hard metal parts, this seal has a long life and is suitable for a wide range of cutting fluids, including pure water. Another method is to use two shorter clearance seals on the reciprocating ram. By feeding the space between these seals with oil, which is very viscous at high pressures, the leakage is virtually eliminated, although a small percentage of the oil will inevitably find its way into the cutting fluid through the inner seal.

(ii) *Tubing* High pressure tubing used to transport fluid from one system component to another is thick walled, with the ratio of the outside to inside diameter at least 5 and sometimes as high as 10. The tubing may be made from a solid stainless steel wall or a composite wall with stainless steel inside and carbon steel as a jacket. Tubing may be used to constrain fluids at pressures greater than the yield stress of the tube material by the use of a process known as *autofrettaging* or *self-hooping*.

(iii) *Tube-fittings* A metal-to-metal line contact is the usual technique for achieving a fluid seal in high pressure tube fittings, accomplished by drawing a cone shape into a rounded socket. The cone may be machined directly onto the tubing or a cone shaped insert may be used. At the highest pressures, the replaceable cone design is the most workable.

(iv) *Valves* Most high pressure valves are of the needle type. The main flow passage is controlled by a cone shape on the end of the needle fixed into a seat. A gland seal is usually required to eliminate leaks along the stem.

(v) *Nozzles* Nozzles meant to convert the high pressure liquid to a high velocity jet present a severe challenge to the designer. For minimum erosion, the nozzle material should be extremely hard. Yet, to allow the formation of a smooth contour, the material should be ductile and easily machinable. Nozzles made of sintered diamond or sapphire are available and they can be used as inserts in a steel holder that provides the needed strength and ductility. Diamond, tungsten carbide and special steels have also been used successfully for making quality nozzles. Figure 2.25 shows a nozzle assembly.

A nozzle with an exit diameter of 0.05-0.35 mm gives a coherent jet length of up to 3-4 cm. A method of increasing this length is to add to the cutting water up to 1 per cent of a long chain polymer, such as polyethylene oxide with a molecular weight of four million, which produces much higher fluid viscosities. With such an additive, coherent jets of up to

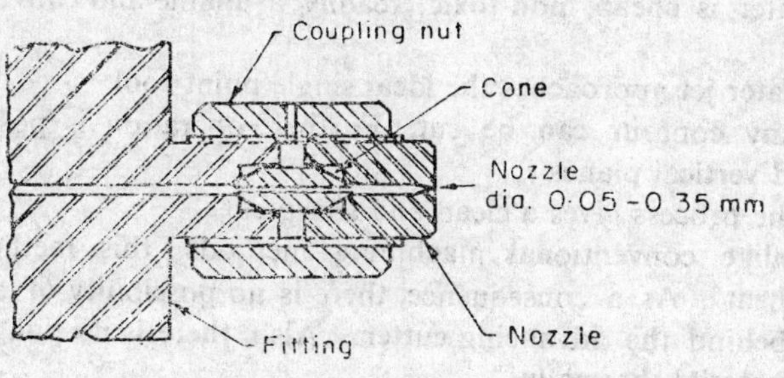

Fig. 2.25

600 diameters in length have been achieved. Even beyond the break-up point, some cutting action is still possible as a concentrated liquid core remains within the growing spray envelope.

Process Details

The water and polymer are mixed properly and the mixture is sent to the intensifier where its pressure is raised. (As mentioned earlier, the function of polymer is to check the divergence of the stream coming from the nozzle.) The hydraulic intensifier increases the intensity of the pressure of water and supplies it to a hydraulic accumulator (reservoir) since the energy is not required continuously. During the idle periods, energy is stored in the accumulator and given out during cutting. The intensifier, therefore, need not supply as high an amount of energy as is required by the cutting process when doing maximum cutting.

The high pressure water coming from the accumulator is controlled by a control panel from where it goes to the nozzle after passing through the stop-start valve. The jet stream coming out of the nozzle cuts the work piece and is then collected in a drain system.

Figure 2.26 shows the block diagram of a jet cutting system.

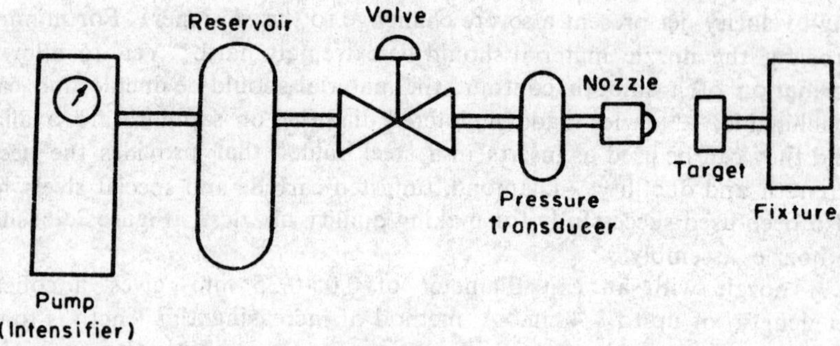

Fig. 2.26

Advantages

The advantages of this process are:

(i) Water is cheap, non-toxic, readily available and can be easily disposed.

(ii) Water jet approaches the ideal single point tool.

(iii) Any contour can be cut. Further operation is possible in horizontal and vertical planes.

(iv) The process gives a clean and sharp cut.

(v) Unlike conventional machining methods, this method does not generate heat. As a consequence, there is no possibility of rewelding the material behind the advancing cutter. Also, there is no danger of degrading the material thermally.

(vi) Best suited for explosive environments.

(vii) Dustless atmosphere—this is particularly advantageous for cutting asbestos and glass fibre insulation materials which produce dust.

(viii) Noise is minimized as the power units and pumps can be kept away from the cutting point.

(ix) No moving parts are present and, therefore, less maintenance is required.

(x) Jet takes away all the cutting residue and hence there are on pollution problems.

(xi) Fluid can be re-used by filtering out the solids.

(xii) Only a small amount of fluid is required (usual requirement is of the order of 100-150 litres/hour).

Practical Applications

Hydraulic mining of coal has been carried out in Russia, China, Poland, Czechoslovakia, Canada and Germany, while experiments have been conducted in other countries, including Britain.   In most mining applications water pressures below 3.5 kN/cm^2 combined with large diameter nozzles have been used.   The large amount of water used also serves to carry the broken coal from the face.   A Canadian mine reported that two men had produced 2250 tonnes of coal in one shift by hydraulic mining.

Current developments are aimed at combining water jets with mechanical methods of cutting both coal and rock.   The high pressure jets cut slots which weaken the material, making it an easy task for relatively conventional cutters to break the coal or rock.   For example, ESSO used conventional roller drilling bits to which jets were added in some of their oil-well drilling tests.   A two-to three-fold increase in drilling speeds were achieved with six nozzles, each 3.3 mm in diameter, operating at a pressure of 7-10 kN/cm^2 in conjunction with a 222 mm diameter drilling bit.

Japanese railway engineers have tested percussive drills incorporating two to four nozzles, with diameters of 0.2-0.4 mm, operating at 40 kN/cm^2 to achieve holes 35-215 mm in diameter at drilling speeds between two to five times the normal rates.   The Japanese interest in this method is to help speed up tunnel boring so as to extend their high speed railway network.   Similar methods have been used for demolishing reinforced concrete structures, cutting anti-skid grooves in airfield runways and roads, making trenches, and laying cable.

The other area in which the destructive power of water jets can be applied is cleaning and descaling.   Hand-held lances operating at pressures of 3-10 kN/cm^2 are widely used in oil refineries, chemical plants, power stations, steel works and dry docks.   Such jets can be lethal and it is important that trained personnel should wear protective clothing and follow strict operational procedures.

Jet cutting is particularly suitable for difficult-to-machine materials, such as 25 mm aluminium honeycomb, which can be cut at speeds of 0.85 m/s or newsprint at 30 m/s. When it comes to harder materials, Japanese

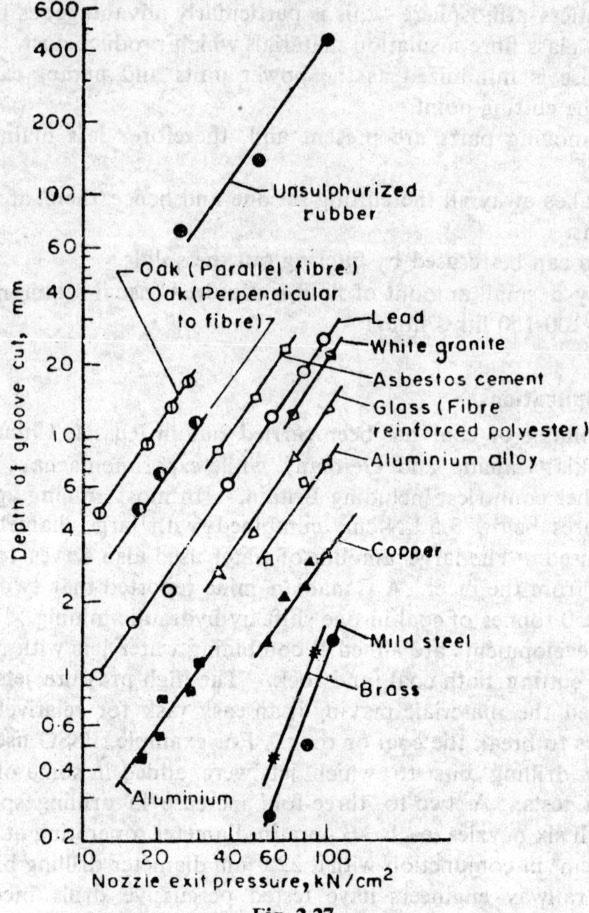

Fig. 2.27

engineers have published some interesting results (Fig. 2.27) which indicate that even steel and brass can be cut, although the pressures needed are as yet beyond the range of the commercial intensifiers that are currently available.

Today, a large number of industries are either using or planning to use jet cutting, including board and paper products, food, asbestos and footwear. Control systems have become more sophisticated since the first installation was commissioned in the USA in 1971. Armchair shapes were achieved using templates and some manual movement; now the same company has an optical tracing system working from a drawing to control multiple nozzles. Numerical control will be the next step, enabling many different shapes and sizes of products to be cut.

REVIEW QUESTIONS

1. Define 'ultrasonics' and describe the process in which these are used to machine the material.
2. What are the constituents of slurry used in ultrasonic machining? Name the characteristics of a good suspension media. Which fluid satisfies most of these requirements?
3. Sketch and describe any two types of tool feed systems used in ultrasonic machining.
4. Discuss the hypothesis proposed by Shaw regarding the mode of material removal in ultrasonic machining and obtain an expression for machining rate. What are the assumptions on which this expression is based? How far are these assumptions valid?
5. Discuss the effects of the following parameters on the rate of material removal and surface finish obtainable in ultrasonic machining:
   (i) Amplitude and frequency of vibration.
   (ii) Abrasive grit size.
   (iii) Static load.
6. One of the limitations of ultrasonic machining process is that the cutting rate obtained is not competitive. Write a note on the studies made and the results achieved to overcome this limitation.
7. Calculate the depth of indentation produced on a glass surface in ultrasonic machining by the throwing action of abrasive grain of 100 μm diameter. The following data are available:

Amplitude of vibration	0.1 mm
Frequency	20 kc/s
Abrasive density	3.0 kg/m³
Yield strength of glass	4.0×10^{11} N/m²

Outline a method by which the volume rate of material removal could be computed.
8. Discuss why the AJM technique, when applied to ductile materials, leads to a low rate of metal removal.
9. Discuss the effects of the following parameters on working accuracy and rate of metal removal in AJM:
   (i) Grain size.
   (ii) Jet velocity.
   (iii) Stand off distance.
10. Describe at least three typical engineering applications of AJM.
11. (i) What is the principle of water jet machining?
    (ii) Write a note on the special features of the equipment used in this method of machining.
    (iii) Give the practical applications of water jet machining.

3

Electrochemical and Chemical Metal Removal Processes

Electrochemical machining (sometimes referred to as anodic cutting) is one of the latest and potentially the most useful of the non-traditional machining processes. The basic principles of the process are not new but applications of the process as a metal working tool are definitely new. Extensive development of the process has taken place in recent years mainly due to (i) the need to machine harder and tougher materials, (ii) the increasing cost of manual labour, and (iii) the need to machine configurations beyond the capability of conventional machining methods.

The removal of material from selective areas by using chemicals is also an old art. Some advances have been made in this technology in the past few years and many new applications have been tried.

In this chapter, various electrochemical and chemical processes are described which are commercially significant.

ELECTROCHEMICAL MACHINING (ECM)

Michael Faraday discovered that if two electrodes are placed in a bath containing a conductive liquid and D.C. potential is applied across them, metal can be deplated from the anode and plated on the cathode. This principle was in use for a long time in a process called 'electroplating'. With certain modifications, ECM is the reverse of electroplating.

ECM uses a shaped tool or electrode. Since the term 'machining' implies the removal of material from the work piece, the tool is made the cathode and the work piece the anode. An electrolyte is pumped through the small gap which is maintained between the tool and work piece. The chemical properties of this electrolyte are such that the constituents of the work material go into the solution by the electrolytic process but do not plate on the tool. By designing the tool accurately, the cavity or hole which is produced exactly reproduces the tool shape. Figure 3.1 shows the schematic representation of the principles of electrochemical machining.

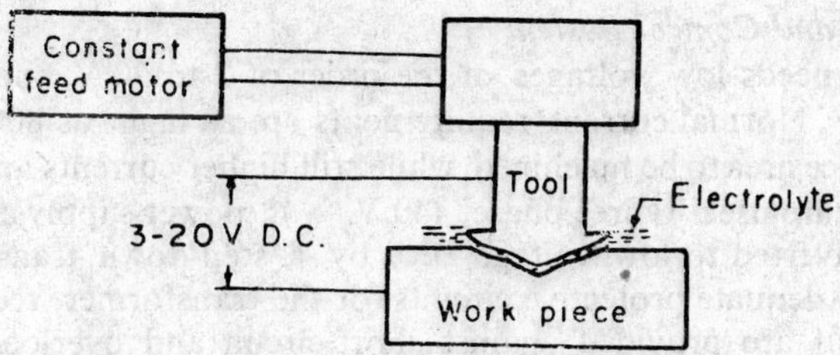

Fig. 3.1 Schematic of electrochemical machining

Elements of ECM Process

The elements of the process are:

 (i) Cathode tool (of a shape which is almost the mirror image of the cavity to be machined into the work piece).
 (ii) Anode work piece (and means to hold and locate it near the tool).
 (iii) Source of D.C. power (of sufficient capacity so that high current densities can be maintained between the tool and work piece).
 (iv) Electrolyte, a conductive liquid (and means to supply it into the gap between the tool and work piece).

Each of these elements are discussed in detail.

Cathode Tool

The accuracy of the tool shape directly affects the work piece accuracy, since configuration of the cavity produced cannot be more accurate than the tool that produces it. The same is applicable to the surface finish of the tool. Poor surface finish of the tool will produce a poor surface on the part.

The materials that find wide applications in the manufacture of tools for ECM are aluminium, brass, bronze, copper, carbon, stainless steel, monel and reinforced plastics.

Shaping the tool is usually not a problem since all these materials are easy to machine. Cold forging and electroforming are favourable methods of tool making.

Anode Work piece

There is no restriction on the nature of the work material except that it must be a good conductor of electricity. The chemical characteristics of the work material, however, do affect the material removal rate. The removal rate is proportional to the atomic weight and inverse of the valency of work material.

The fixtures for holding the work are made of some insulating material. Epoxy resins or glass fibre resins have been successfully employed for this purpose. Cheaper materials which serve the purpose in some instances are perspex and PVC. They should have good thermal stability and low moisture absorption properties. Some provision is made through the fixture for the electric contact between the supply and the work piece.

D.C. Power and Control System

The process needs low voltages of the order of 2 to 20 V, and in rare cases up to 30 V. Normal current requirements are as high as 800 amp/cm^2 of the work piece area to be machined, while still higher currents are being considered for future use. Three phase, 440 V, A.C. power supply available from mains is converted to low voltage D.C. by a step-down transformer and a rectifier. Adequate protective circuits for the transformer, rectifier and the machine itself are provided against short-circuit and overload conditions. These conditions may occur due to the mishandling or incorrect fitting of electrodes or work piece, piling up of conducting debris in the gap or malfunctioning of the gap-control system.

A schematic diagram of the system of power supply for an ECM machine is shown in Fig. 3.2.

Several methods are available for achieving a constant output voltage. One of these methods employs a motorized variable transformer which varies the input to the step-down transformer in such a way as to maintain D.C. output at a constant pre-set voltage. Another method, which is superior to the first one, employs a saturable reactor to control the output. This is connected in series with the primary winding of the step-down transformer, and the currents in the control windings are so varied that the D.C. output remains at the pre-set voltage. The use of silicon-controlled rectifiers is becoming very common for voltage control in ECM machines. These are light in weight and compact in size.

Owing to loss or lack of control, sparking may occur between the tool and the work piece, causing damage to both. The damage increases with the duration of the spark. Preventive measures for sparking are necessary if continuous production of good quality parts is to be obtained. Several electronic sensing devices are available for detecting faults that may occur within the electrodes gap.

Electrolyte

The electrolyte used in ECM performs many functions, such as:
  (i) Completing the electric circuit between the tool and the work piece.
 (ii) Allowing desirable machining reactions to occur.
(iii) Carrying away heat generated during the chemical reactions.
(iv) Carrying away products of reaction from the zone of machining.

Essential properties and selection An effective and efficient electrolyte should have the following characteristics:
  (i) High electrical conductivity.

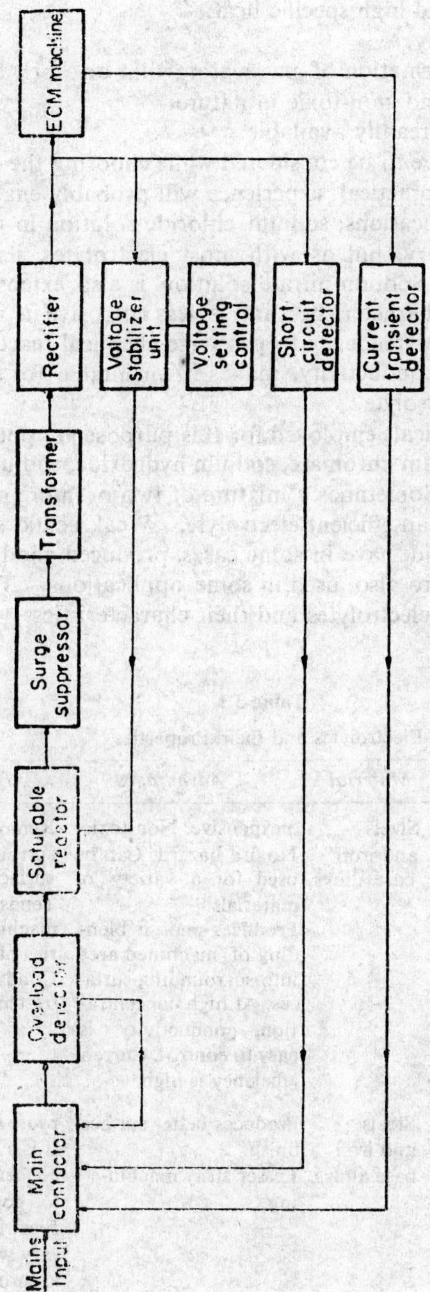

Fig. 3.2 System of power supply for ECM machine

(ii) Low viscosity and high specific heat.

(iii) Chemical stability.

(iv) Resistance to formation of passivating film on work surface.

(v) Non-corrosive and non-toxic in nature.

(vi) Inexpensive and readily available.

All these factors have to be considered while choosing the best electrolyte for a specific use and practical experience will probably enable the best selection. In most applications, sodium chloride solution in water has been found to be satisfactory, but as with most electrolytes, its corrosiveness presents a problem. Sodium nitrate solution is also extensively used. It has all the desirable characteristics and is less corrosive in nature than sodium chloride. But its tendency to passivate chemical reaction and somewhat lower electrical conductivity, makes it unsuitable for adoption as an all purpose ECM electrolyte.

Other alkaline chemicals employed for this purpose are potassium nitrate, sodium sulphate, sodium chromate, sodium hydroxide, sodium fluoride and potassium chloride. Sometimes a mixture of two or more of these chemicals is best suited for an efficient electrolyte. Weak acidic solutions, such as that of sulphuric acid, have in some cases, produced good surface finish. Neutral electrolytes are also used in some applications. Table 3.1 gives some commonly used electrolytes and their characteristics.

Table 3.1

Electrolytes and their Properties

S. No.	Electrolyte	Material	Advantages	Disadvantages
(i)	NaCl or KCl upto 0.25 kg/ litre.	Steels and iron base alloys	Inexpensive. Non-toxic. No fire hazard. Can be used for a variety of materials. Produces smooth blending of machined area into surrounding surfaces. At high concentration, conductivity is easy to control. Current efficiency is high.	Removes material from surrounding finished surfaces. Tends to selectively machine grain boundaries of stainless steels. Conductivity is affected by temperature.
(ii)	NaNO₃ upto 0.5 kg/litre.	Steels and iron base alloys.	Produces better surface finish. Lesser stray machining.	More costly than NaCl. It is a fire hazard. High concentration is required for good conductivity. Electrical resistivity can vary with flow velocity to produce flow lines. Higher voltages are required. Conductivity is affected by temperature.

(Contd.)

Table 3.1 (*Contd.*)

S. No.	Electrolyte	Material	Advantages	Disadvantages
(iii)	$NaClO_3$ 0.2 to 0.5 kg/litre.	Steel and iron base alloys.	Produces very smooth bright surface which is resistant to subsequent corrosion. Conductivity relatively unaffected by temperature. No stray machining or pitting of adjacent surfaces. Applicable for precise machining.	High cost. Toxic. High fire risk if allowed to dry on combustible materials. Requires special handling and storage. Low current efficiency at lower concentration. High voltages (about 20 V) are required for best results. No machining at <9 V.
(iv)	$NaCl+0.01$ kg/litre citric acid.	Steel and iron base alloys.	Citric acid prevents formation of metal hydroxide precipitates. Suitable for operation at very small gap sizes for definition of intricate work shapes. Lower electrolyte pressures may be employed, e.g. in the operation of slender electrodes as used in deep hole drilling.	Very expensive if used for other than small intricate parts. Electrolyte must be changed frequently.
(v)	$NaCl$ upto 0.20 kg/litre or $NaNO_3$ upto 0.5 kg/litre.	Grey cast iron.	Produces good surface finish.	Large machining gap sizes must be used (0.50-0.80 mm) to allow graphite particles to be swept clear.
(vi)	$NaCl$ upto 0.1 kg/litre + $NaNO_3$ 0.1 kg/litre.	Nickel and cobalt-base alloys.	Inexpensive mixture. Produces good surface finish.	Produces flow lines on work surface. Machined surface carries lightly adherent grey smut.
(vii)	$NaNO_3$ upto 0.5 kg/litre.	Nickel and cabalt-base alloys.	Produces good surface finish.	Disadvantages same as (vi) electrolyte. Costlier than (vi) electrolyte. It is a fire hazard. Needs higher operating voltage.
(vii)	$NaCl$ upto 0.25 kg/litre.	Nickel and cabalt-base alloys.	Produces smooth surfaces. Surfaces need no after cleaning.	Intergranular corrosion which drastically lowers the fatigue endurance of components machined with this electrolyte. Danger of pitting

(Contd.)

Table 3.1 (*contd.*)

S. No.	Electrolyte	Material	Advantages	Disadvantages
				of surfaces adjacent to machined area.
(ix)	NaCl upto 0.25 kg/litre+0.002 kg/litre NaF.	Nickel and cabalt-base alloys.	Materials which display irregular black oxide films when machined with any of the above electrolytes will machine to a good finish with this electrolyte.	Same disadvantages as listed for NaCl.
(x)	NaCl upto 0.15 kg/litre+0.05 kg/litre NaBr+ 0.002 kg/litre NaF.	Titanium alloys.	Produces good surface finish. Characteristics of surface oxide films insensitive to flow velocity. Easy to control contour dimensions.	Higher voltage reduces chance of passivity. A loose grey oxide film remains on work surface. Danger of pitting on surface adjacent to machining area.
(xi)	NaCl less than 0.1 kg/litre.	Titanium alloys.	Can produce surface finishes better than previous electrolyte. Machined surface does not have any loose oxide film.	High voltages are required. Flow lines may be formed on the work surface. Surface is more prone to passivity during machining. Close contour dimensions are difficult to maintain. Pitting occurs on surfaces adjacent to machining area.

Concentration   A concentrated electrolyte offers low resistance to the flow of current. A greater density is achieved for a specific operating voltage. But the disadvantage is that salts crystallize out of the solution at higher concentrations and clog the areas in the machine-work enclosure.

Dilute electrolytes are useful in certain cases. If, for example, work surface finish is important, high machining voltage, dilute electrolyte and a small gap should be employed instead of a low voltage and concentrated electrolyte.

Working life   The composition of the electrolyte changes during use. The main changes that occur and their effects are given below:

(1) Loss of hydrogen. This reduces the electrical conductivity of the electrolyte.

(2) Loss of water due to evaporation or being carried off by the hydrogen gas generated. Concentration of the solution increases which affects viscosity and electrical conductivity.

(3) Precipitate formation. This increases the effective viscosity and interferes with the process in the work-electrode gap.

(4) Salt absorption by the precipitate. This reduces concentration and affects electrical conductivity.

These changes indicate that the electrolyte has a finite working life. In practice, its life may be limited due to the difficulty in maintaining a reasonably constant electrical conductivity for control of the process and accuracy in machining, and in avoiding precipitation.

Temperature and pressure The difference in the temperatures of the electrolyte at the entrance and exit of the tool-work gap is an important factor. In their experimental work, Kubeth and Heitmann [53] recorded a temperature difference of as much as 45° C under otherwise normal process parameters. Such a temperature difference can be expected to decrease the specific resistance of the 10 per cent NaCl electrolyte by about 50 per cent. These changes would definitely affect the flow rate/pressure characteristics of the electrolyte. Precautions are taken to prevent the overheating of the electrolyte.

There are many advantages of using hot electrolyte too. An increase in the temperature speeds up electrode reactions and reduces the over-voltages required. As shown in Fig. 3.3, the voltage (and thus the power) required to maintain a given current density is reduced with a rise in temperature. For hot electrolyte, the pressure required to force it through the gap at the given rate is low. The solubility of the products of reaction increases with rising electrolyte temperature.

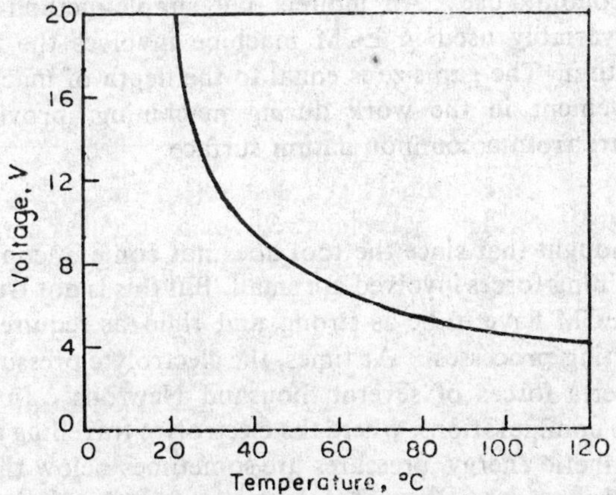

Fig. 3.3 Variation of machining voltage with electrolyte temperature

In order to obtain good results, it is essential to (i) remove the contaminated electrolyte from the working gap, (ii) minimize polarization, and (iii) limit the rise in temperature of the electrolyte, particularly when working at higher current densities. Increased pressure of electrolyte above atmospheric pressure also increases the boiling point of the electrolyte, reduces the over-voltage at the tool surface and decreases the volume of generated hydrogen. Hydrogen bubbles thus occupy a smaller volume of the gap and displace less electrolyte, permitting the use of higher current densities.

Tool-Work Gap

The controllable parameters which determine the gap size are electrolyte temperature and composition, voltage, and tool feed rate. Those relating to the electrolyte are normally not difficult to maintain at constant values. The gap can, therefore, be adjusted by voltage and tool feed rate. Under equilibrium conditions, the tool feed rate (or the material removal rate) is directly proportional to current density. If the feed rate is increased, the electrical resistance of the gap reduces to allow more current to flow. From this it follows that if the feed rate is doubled, the gap will be halved. While it holds for some metals, there is additional resistive loss that occurs at the tool and work surfaces. The voltage loss at the work surface is termed as 'anode potential'. The gap size thus becomes less than half in reducing the total resistive path to half to allow the tool feed rate to be doubled. In the case of voltage variations also, the same reasoning holds good. The change in the gap is more than the proportionate change in voltage.

No satisfactory methods are currently available for gap measurement while the process is in operation. Those employing lasers or electronic monitoring of current with imposed tool vibrations are cumbersome for practical and economic use. An indirect but simple method of measurement that is invariably used in ECM machine involves the tool position from a fixed datum. The gap size is equal to the depth of machining minus total tool movement in the work during machining, provided that all measurements are from a common datum surface.

ECM Machine

It was earlier thought that since the tool does not come in contact with the work, the machining forces involved are small. But this is not true. Machines and tooling in ECM have to be as strong and rigid as required in conventional metal cutting processes. At times, the electrolyte pressure within the working gap exerts forces of several thousand Newtons. In the case of certain electrode configurations, where the electrolyte travelling at high speed possesses only kinetic energy, pressures are sometimes below that of the atmosphere, and there is a tendency for the tool to be drawn into the work piece. The structure of machines must be strong and rigid to limit the deflections.

ECM machines are available in many types and sizes, having different systems for loading, set-up, tooling, alignment between work and tool and methods for control. A typical ECM machine may have a table for loading the work and a platen fitted to a ram for mounting the tool. Provisions are made for efficient electrical contact of the work piece with the positive terminal, and of the tool with the negative terminal of the power supply. A pump supplies the electrolyte under pressure (about 1.5 N/mm^2) into the gap between the work and tool. Hydrogen generated in the process is removed from the work enclosure to prevent any serious explosion.

Modern ECM machines are equipped with the means to safely handle high currents as well as special drive systems that ensure slow and accurate movement of the tool. They are also lined with non-corrosive materials for protection from the corrosive action of electrolytes.

Chemistry of the Process

In the electrolytic circuit shown in Fig. 3.4, the direction of electron flow is from the work piece through the power supply to the tool. Since electrons

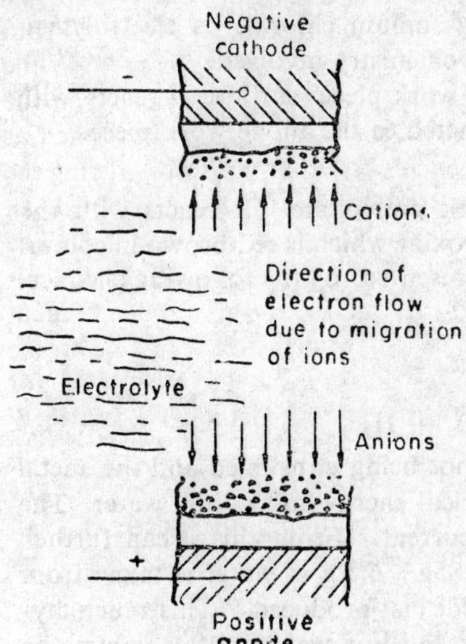

Fig. 3.4 Electrolytic circuit for electro-chemical reactions in electro-chemical machining

cannot flow through the electrolyte, the electric current is maintained by the removal of the electrons from the atomic structure of the work piece. The least strongly bound electrons are found at the surface of the work piece. These electrons dissociate themselves from the work piece and flow into the electric circuit.

Many chemical reactions occur at the cathode, the anode and in the electrolyte. At the tool (cathode), the reaction having the smallest oxidation potential will take place, and at the work piece (anode), the reaction having the largest oxidation potential will occur first.

At the cathode, the following reactions are possible:

$$M^+ + e^- \rightarrow M \text{ (Metal)}.$$

$$2H^+ + 2e^- \rightarrow H_2 \text{ (Hydrogen evolution)}.$$

The factors which influence the oxidation potential, and thus determine which of these reactions will occur, are:

(i) Nature of metal being machined.

   (ii) Type of electrolyte.
  (iii) Current density.
  (iv) Temperature of electrolyte.

Halide salts which are commonly used as electrolytes, give rise to simple electrolytic reactions. The following reactions occur at the anode with a halogen electrolyte:

$$M \rightarrow M^+ + e \text{ (Metal dissolution).}$$
$$2H_2O \rightarrow O_2 + 4H^+ + 4e \text{ (Oxygen evolution).}$$
$$2Cl^- \rightarrow Cl_2 + 2e \text{ (Halogen gas evolution).}$$

It has been established that the metal dissolution reaction is the main or the only reaction that occurs. When the metal ions leave the work piece surface, many reactions occur in the electrolyte. A typical example of the electrochemical machining of iron using sodium chloride as electrolyte is given here, which is representative of the chemistry involved.

As an ion of iron (Fe^{++}) leaves the work piece surface, it reacts with hydroxyl ions (OH^-) that have been attracted to the anode work piece.

$$Fe^{++} + 2(OH^-) \rightarrow Fe\,(OH)_2$$

This ferrous hydroxide is a green-black precipitate. It reacts with the air of the atmosphere to form ferric hydroxide which is red-brown in colour.

The complex ECM process can be represented by the following chemical equation:

$$Fe + 2H_2O \rightarrow Fe\,(OH)_2 + H_2$$

  or

$$2Fe + 4H_2O + O_2 \rightarrow 2Fe\,(OH)_3 + H_2$$

It is interesting to note that the salt is not being consumed and the metal is being machined at the expense of electrical energy and a little water. The electrolyte acts only as a carrier of the current. From this it can further be seen that in removing 1 cm³ of iron (7.85g), 6.3 g of water is taken from the electrolyte and 15 g of ferric hydroxide is produced. This much hydroxide has a volume of 4 cm³ when dry, but in a wet state the volume of sludge would be about 300 cm³. Of the 6.3 g of water, 0.28 g goes as hydrogen gas. In other words, a current of 1000 A would dissolve iron at the rate of about 15 g/min and generate hydrogen at the rate of about 600 cc/min.

Metal Removal Rate

Theoretically, the metal removal rate can be estimated as follows:

      Referring to Fig. 3.5, let

      h—'tool-work' gap (m)

      A—area of the current path (m²)

      ρ_s—specific resistance of electrolyte (ohm-m)

      E—machining voltage (volt)

      I—current flowing through gap (amp)

      t—time for which current flows (s)

      Current density (amp/m²) S is given by

$$S = \frac{\text{Voltage}}{\text{Resistance} \times \text{Area}} = \frac{E}{\dfrac{\rho_s h}{A} \cdot A} = \frac{E}{\rho_s h} \qquad (3.1)$$

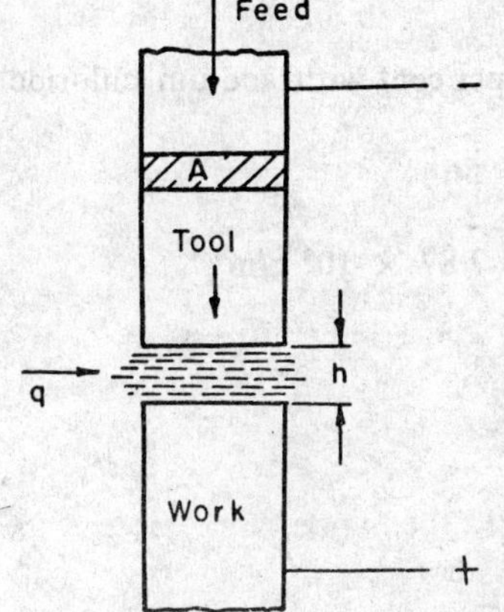

Fig. 3.5 A model for determining material removal rate in electrochemical machining

According to Faraday's first law of electrolysis, the chemical change produced during electrolysis is proportional to the quantity of the electricity passed and to the electrochemical equivalent of the anode (work) material, that is,

$$m \propto I \cdot t \cdot (N/n)$$
$$= (1/96500)\, I \cdot t \cdot (N/n) \qquad (3.2)$$

Volume of metal removed by any quantity of electricity

$$= \frac{1}{96500}\, I \cdot t \cdot \frac{N}{n} \cdot \frac{1}{d} \cdot \eta$$

where N is the atomic weight of work material, n the valency of work material, d the density of work material, and η the current efficiency which may be defined as the efficiency of the current in removing metal from the work piece. The current efficiency is close to 100 per cent when sodium chloride is used as the electrolyte, but for nitrate and sulphate solutions, it is somewhat lower.

Specific metal removal rate (s) can be written as

$$s = \frac{1}{96500} \cdot \frac{N}{n} \cdot \frac{1}{d} \cdot \eta \ (\text{m}^3/\text{amp s}) \qquad (3.3)$$

and feed rate of electrode

$$f = S \times s$$
$$= [E/\rho_s h]\,[N/n]\,[1/d]\,[1/96500]\ (\text{m/s}) \qquad (3.4)$$

EXAMPLE

Calculate the machining rate and the electrode feed rate when iron is electrochemically machined, using copper electrode and sodium chloride solution (specific resistance = 5.0 ohm cm). The power supply data of the ECM machine used are:

Supply voltage 18 V D.C.

> Current   5000 amp
> A 'tool-work' gap of 0.5 mm (constant) may be assumed.

SOLUTION

The current efficiency η can be taken as 100 per cent with sodium chloride electrolyte.

For iron (anode), atomic weight $N = 56$
$$\text{valency } n = 2$$
$$\text{density } d = 7.87 \times 10^6 \text{ g/m}^3$$

Specific metal removal rate

$$s = \frac{1}{96500} \cdot \frac{N}{n} \cdot \frac{1}{d}$$

$$= \frac{1}{96500} \cdot \frac{56}{2} \cdot \frac{1}{7.87 \times 10^6}$$

$$= 3.67 \times 10^{-11} \text{ m}^3/\text{amp/s}$$

Metal removal rate $= s \cdot I$

$$= 3.67 \times 10^{-11} \times 5000$$
$$= 1.835 \times 10^{-7} \text{ m}^3/\text{s}$$

Electrode feed rate $= \dfrac{E}{\rho_s h} \cdot s$

$$= \frac{18 \times 3.67 \times 10^{-11}}{0.05 \times 5 \times 10^{-4}} \text{ m/s}$$

$$= 1.585 \text{ mm/min}$$

A study of Eqs. 3.3 and 3.4 indicates that if the voltage, feed rate and resistivity of the electrolyte are kept constant, a uniform gap will exist and absolute conformity to the tool shape will be obtained. But it is not so easy to analyze the process or predict the results. It is not possible to maintain constant electrolyte resistivity in the gap because of the influence of many parameters. Electrolyte temperature, due to heat generated during chemical reactions, tends to reduce resistivity. The evolution of hydrogen gas and any flow disturbances, as the electrolyte progresses through the gap, also affect the electrolyte resistivity. In practice, the process is further complicated by the presence of a polarized ion layer at either or both electrodes. The temperature rise reduces the viscosity and, therefore, also the 'pressure-flow rate' relationship.

It is known that current density and field strength tend to be higher at sharp edges and corners. This results in non-uniform gaps because of higher metal removal rates in those areas, and that is why it is difficult to machine sharp internal corners by this process.

It is interesting to note the 'self adjusting' feature of the ECM process. If the tool advances at a rate faster than that of metal removal, the gap becomes narrower, current density increases and consequently the machining rate also increases. Within a short time, a steady gap is established, satisfying the relation, $h = E \cdot s/\rho_s f$. Conversely, if the feed rate is lower than the material removal rate, the gap will widen for a short time and the current density will fall due to higher resistance, and again, the machining rate is so established that it matches the feed rate.

Theoretically, there seems to be no upper limit to the rate of penetration of the electrode. However, the only parameter that restricts an infinite penetration rate is the capacity of the electrolyte flowing through the gap to carry away heat, without causing boiling. If it is assumed that the rise in temperature of the electrolyte is purely due to the I^2R loss in the electrolyte, and that in the extreme case the temperature rise of the electrolyte can be permitted up to boiling point, an approximate expression for the maximum allowable rate of penetration can be derived.

Referring to Fig. 3.5, let

C —specific heat of electrolyte (cal/kg/K)

ρ_e —density of electrolyte (kg/m³)

f_p —maximum permissible feed (m/s)

I_p —maximum permissible current (amp)

m_e —mass of electrolyte passing through the gap (kg)

q —volume rate of flow of electrolyte (m³/s)

T_b —boiling point of the electrolyte (K)

T_1 —inlet temperature of the electrolyte (K)

H —heat to raise m_e kg of electrolyte from T_1 to T_b K (cal)

P —power equivalent of H, (W)

t —time for which current flows (s)

Then

$$H = m_e\, C\, (T_b - T_1)$$
$$= V_e \rho_e C (T_b - T_1)$$

$$\frac{H}{t} = \frac{V_e}{t} \cdot \rho_e\, C\, (T_b - T_1)$$

or

$$P = 4.186\, q\, \rho_e\, C\, (T_b - T_1)$$

or

$$I_p^2 R = 4.186\, q\, \rho_e\, C(T_b - T_1)$$

or

$$I_p = \sqrt{\frac{4.186\, q\, \rho_e\, C\, (T_b - T_1)}{R}} \tag{3.5}$$

But

$$R = \frac{\rho_s h}{A}$$

$$I_p = \sqrt{\frac{4.186\, q\, \rho_e\, CA\, (T_b - T_1)}{\rho_s \cdot h}} \tag{3.6}$$

Maximum permissible feed rate of the electrode

$$f_p = S \times s$$
$$= \frac{I_p}{A} s$$

$$= \frac{1}{96500} \cdot \frac{N}{n} \cdot \frac{1}{d} \cdot \eta \sqrt{\frac{4.186\, q\, \rho_e\, C\, (T_b - T_1)}{\rho_s \cdot h \cdot A}}$$

$$= \frac{1}{47200} \cdot \frac{N\eta}{n\,d} \sqrt{\frac{q\, \rho_e\, C\, (T_b - T_1)}{\rho_s\, h\, A}} \tag{3.7}$$

This equation can also be used to calculate the electrolyte temperature change for any desired electrode feed rate, as

$$\Delta T = 2.23 \times 10^9 \left[\frac{\rho_s\, h\, A}{q\, \rho_e\, C}\right] \left[\frac{f\, n\, d}{N\, \eta}\right]^2 \tag{3.8}$$

Discrepancies are sometimes observed between the metal removal rates and electrode feed rates estimated by Eq. 3.4 and the actual experimental results. In practice, metal removal rates are often higher than the estimated ones. The probable reasons for this are:

(i) The exact valency at which a metal behaves in the electrochemical reaction is generally unknown. Some metals have only one valency while others have many; the valency at which they enter the reaction is not known. Chromium and nickel are metals for which exact metal removal rates cannot be estimated unless valency for these are accurately known.

(ii) Electrochemical machining continuously exposes a new and clean surface to the electrolyte which is easily attacked chemically. The extent of 'chemical machining' varies, depending upon the electrolyte used and the work metal machined. Metals, such as aluminium are easily attacked and quite a good proportion of chemical machining can occur. This chemical machining adds up to give a higher metal removal rate than the theoretical one.

Tool Design

Tools and fixtures are required to operate for long periods in a corrosive environment of electrolyte and stray electric currents. In order to avoid the rapid corrosion of each tool, the selection of proper material is very important. Generally, stainless steel, copper, brass, bronze, monel, reinforced plastics or copper-tungsten alloy are used. Parts that have anode potential corrode rapidly and, therefore, the number of parts in electrical contact with the work piece is limited. Non-metallic materials may be useful as these are electrically non-conductive and chemically corrosion-resistant.

All electrolyte ducts need to be made of non-corrosive materials, as one of the major process requirements is that no particles of corrosion should enter the electrolyte flow in the tool-work gap. To prevent overheating, there is a limit to the minimum cross-section of the current-carrying parts. For 1000 amp, it is about 6 cm² for copper, 25 cm² for bronze and brass, and 250 cm² for stainless steel. Smaller areas are permissible for metal surfaces cooled by rapidly flowing electrolyte.

Fixtures and tools should be rigid enough to avoid vibration or deflection under the high hydraulic forces which they are subjected to. Proper alignment between the tool and work fixture is essential and is best achieved with removable setting pieces.

Electrical joints should be strictly limited as these are sources of power loss and at times may fail under the wet corrosive conditions in the work enclosure.

Cathode Tool

First, an appraisal should be made of the component to be electrochemically machined. This includes considerations of the economic use of the ECM process, possibility of post treatment to improve the fatigue properties of the component, and other similar factors. The next step usually is to determine the need for any minor change in the component design to simplify ECM tooling. If a number of separate operations are to be performed on the component, the area to be machined by each tool is chosen for the simplest electrolyte flow arrangement and current densities exceeding 15 amp/cm² on all surfaces. In the design of the ECM tool, proper selection of the method of providing adequate electrolyte flow over its working surfaces is of utmost importance. Where bosses or similar feature are to be machined electrolyte can be supplied through a slot in the tool, corresponding to the boss form.

Work corrosion is a problem if electrolyte impinges on adjacent finished work surfaces. This can be avoided by ensuring that a thin protrusion of material is left between adjacent operations which deflects the electrolyte from the work.

Those areas on a tool where ECM action is not required, should be insulated. Lack of insulation on the sides of the die sinking tool (Fig. 3.6) causes unwanted machining of the work and loss in accuracy. Insulation can greatly reduce these effects. The need for a thin insulation can be satisfied with the use of synthetic rubber coating applied in liquid form.

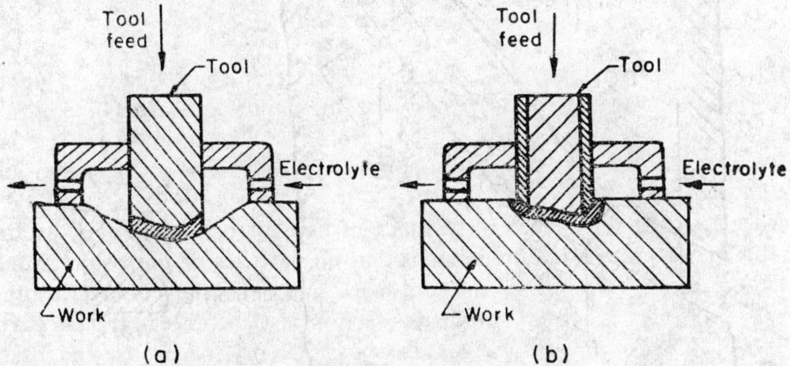

Fig. 3.6 Application of insulating material on areas where ECM action is not
required avoids unwanted machining
(a) A die sinking tool with no insulation on sides
(b) Same tool with insulation on sides

Solid pieces of insulating material secured to the surface should, however, be preferred because they are more reliable for production work.

The shape of the ECM tool is not just the inverse or simple envelope of the shape to be machined. In fact, it bears a complex relationship to the required contour on the work. An approximate method to determine the correction needed in the tool shape is described below.

Correction of Tool Shape

The component shape that is finally obtained in ECM is dependent on the geometry of the tool, the tool feed direction, flow path length and other process variables. The involvement of several factors makes the problem more complex, and for this reason, empirical relations have been determined for tool shape corrections. Proper selection of working parameters can, however, reduce the effect of process phenomena, such as hydrogen formation, so that the problem of tool shape correction is simplified, and analytical methods can be used to derive simple formulae for predicting tool correction. An example is given below of the case of a die sinking ECM tool.

An assumption can be made as to the absence of hydrogen generation, metal precipitation, electrolyte temperature changes in the gap and electric field concentration. Figure 3.7 shows the work profile and the corrected tool profile in dark lines. The dotted line represents the tool shape when the work and tool shapes match exactly and a gap $\angle$ (mea-

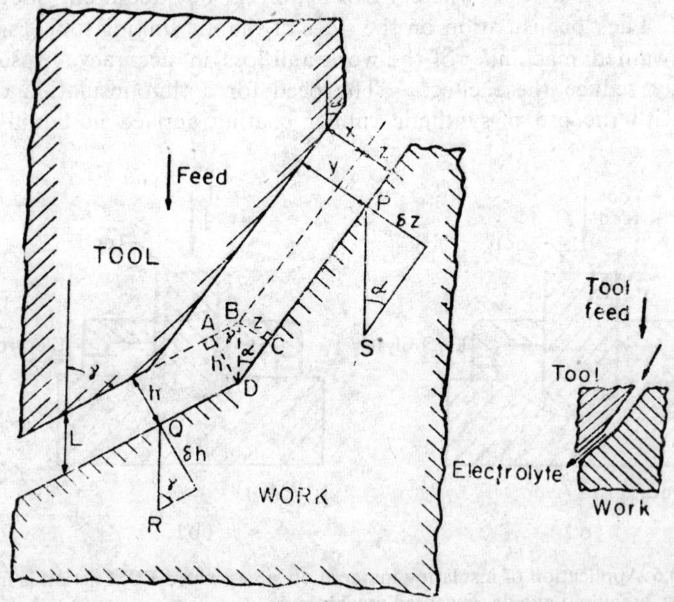

Fig. 3.7 A model of ECM for determining tool profile for given profile of cavity on work surface

sured in the tool feed direction) exists between the tool and work surfaces. The difference between the work and tool shape, at any angle γ on the surface of work, is x.

The minimum tool end gap during operation is h, which occurs at the maximum angle γ of the work surface to the tool feed direction. It follows from Fig. 3.7 that tool correction

$$x = y - z$$

and from triangles ABD snd BCD that

$$\frac{h}{\sin \gamma} = \frac{z}{\sin \alpha}$$

or $$\tag{3.9}$$

$$z = h \sin \alpha \cosec \gamma$$

Taking the metal removal rate as being inversely proportional to the gap size, the incremental metal removal at P and Q can be related as

$$\frac{\delta h}{\delta z} = \frac{y}{h}$$

The metal machining rate in the direction of tool feed at any point on the work surface would be same under equilibrium machining conditions, and thus $PS = QR$, and

$$\frac{\delta h}{\sin \gamma} = \frac{\delta z}{\sin \alpha}$$

or

$$\frac{\delta h}{\delta z} = \frac{\sin \gamma}{\sin \alpha}$$

Therefore

$$\frac{y}{h} = \frac{\sin \gamma}{\sin \alpha}$$

or

$$y = h \sin \gamma \cosec \alpha$$

The tool correction is then

$$x = h (\sin \gamma \cosec \alpha - \sin \alpha \cosec \gamma) \tag{3.10}$$

In most die sinking applications, $\gamma = 90°$ as the tool is fed orthogonally towards some area of the work surface. The above formula can be simplified to

$$x = h (\cosec \alpha - \sin \alpha) \tag{3.11}$$

Accuracy, Surface Finish and other Work Material Characteristics

There are a number of factors which govern the accuracy of the parts produced by ECM. The major ones are

  (i) Machining voltage.

  (ii) Feed rate of electrode (tool).

(iii) Temperature of electrolyte.

(iv) Concentration of electrolyte.

Machining voltage variations severely affect the overcut. A higher voltage brings about a larger overcut and a lower voltage a smaller overcut. In actual practice, the final size of the part is made by controlling the voltage. If a trial cut shows that the overcut should be smaller, voltage is reduced slightly and vice versa. For better accuracy, therefore, there has to be the minimum possible variation in machining voltage. The ECM power supply system attempts to rigidly maintain a constant pre-set voltage for varying incoming line voltage and total machining current.

The feed rate of tool must be maintained under varying conditions of load and machining voltage, otherwise accuracy will be lost. An increase in feed rate, other parameters remaining constant, decreases the overcut and vice versa.

Electrolyte temperature seriously affects the overcut. The power loss (I^2R loss) in the electrolytic reaction gives rise to an increase in the temperature of the electrolyte. This heat must be carried away from the cutting area so as to maintain stable and steady conditions, helpful for achieving better accuracy. To quote an example, in machining a hole 4 cm in diameter $\times$ 30 cm deep, failure to provide adequate electrolyte cooling may cause a 5 K temperature rise in the electrolyte, which is sufficient to increase the gap by 0.2 mm, making the hole approximately half a mm larger in diameter at the bottom than at the top. In general, with proper controlled conditions, a shaping accuracy of $\pm$ 0.0002 mm is common.

The surface finish of a component part depends on its material, the electrolyte and the operating conditions. With nickel-base, cobalt-base and stainless steel alloys, it is nearly 0.1-0.5 microns, and with iron-base alloys and steels, it is 0.5-2 microns. No burrs are formed on machined surfaces.

In general, there is no appreciable change in the mechanical properties, such as tensile strength, yield strength, hardness, ductility, etc. of the material of the parts due to ECM. Values of notched-tensile strength, notched-sensitivity, and the sustained-load characteristics of the ECM parts are comparable with those of conventionally machined parts.

Fatigue strength of stainless steel is found to decrease by ECM. This can, however, be overcome by cold working the surface after ECM.

Economics

Fixed costs of ECM installations are quite high as compared to its operating costs. Overhead costs are the same as for other conventional machining methods. Some costs are unique, such as those of high power, electrode tooling and electrolyte.

As stated in the beginning, ECM needs power of high current capacity. In localities where power is sufficiently cheap, this factor can be overlooked.

Electrode or tooling cost is a fixed cost because there is little wear of the ECM tool. There occurs, however, a negligible abrasion wear of electrode due to electrolyte flow across the gap. With regard to actual tooling cost, it is not very different from conventional machine tooling.

Electrolyte is not as costly as one might think it to be. The most widely used electrolyte is sodium chloride (salt) and it is quite cheap. The normal price of the salt seldom exceeds Re. 0.50 per kg when purchased in large quantities.

Costs of work piece fixtures are not very high. The cost per piece will, however, depend on the number of work pieces finished.

On the shopfloor, ECM installations need not be operated by very skilled engineers and the operation of the machinery can be learnt easily.

The economic success of ECM, in fact, depends largely on the choice of applications. If an operation is simple or if the material can be easily machined by other methods, the high cost of the ECM plant cannot be justified.

An example is considered which compares the ECM process with a conventional machining method on an economic basis.

EXAMPLE

A given component has a complex profile. It is to be produced from a hard nickel-base alloy. The profile can be generated on a copy milling machine or an ECM plant. Presuming that the two machines are fully utilized, determine the number of components for which the cost of component manufacture by the two methods is the same. The following data are provided:

	Copy milling	ECM
Capital cost	Rs. 1,00,000	Rs. 2, 50,000
Amortization period	10 years	5 years
Labour cost/h	Rs. 8,00	Rs. 8,00
Fixture and tooling cost	Rs. 15,000	Rs. 75
Electric power consumption/h	Rs. 2	Rs. 20
Production time/component	40 hours	1 hour

SOLUTION

Assuming 2,500 working hours in a year, the machine costs per hour for copy milling and ECM plant are respectively Rs. 4 and Rs. 20. Further, let there be x number of components, the costs of which on the two machines are equal.

The various costs per component are:

	Copy milling	ECM
Machine time cost	Rs. 160	Rs. 20
Labour time cost	Rs. 320	Rs. 8
Electric powercost	Rs. 80	Rs. 20
Tooling cost	Rs. 15,000/x	Rs. 75,000/x

$$560 + 15,000/x = 48 + 75,000/x$$

or
$$x = 117$$

The break even point is at $x = 117$ components. For total production quantities above this value, ECM is more economical than copy milling (Fig. 3.8). In the above calculations, it is implied that both machines are fully utilized. However, this might not always be the case, particularly for a new process, such as ECM.

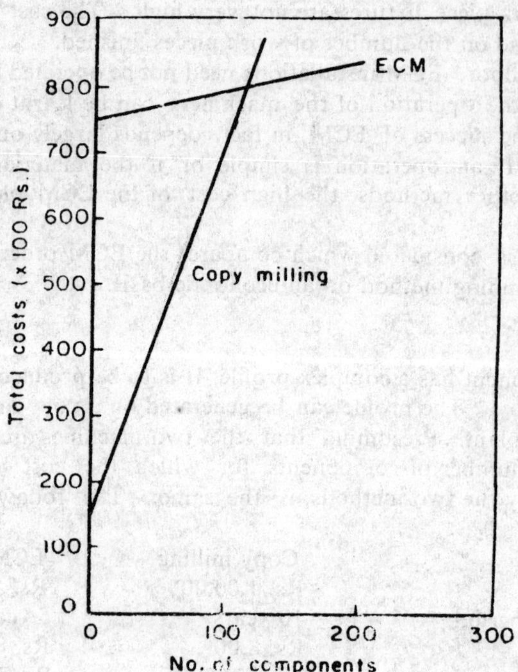

Fig. 3.8 Graphic representation of break-even analysis of ECM and copy milling processes

Advantages

One of the main advantages of ECM lies in its ability to machine complex three-dimensional curved surfaces without the striation marks left by milling cutters. By preparing one set of cathode tools, say, from stainless steel, it is possible to machine many roughly formed components to within close limits of the desired contour. For this reason, gas and steam turbine blades are now machined by this method (Fig. 3.9).

The process is capable of machining metals and alloys irrespective of their strength and hardness Although this advantage is shared with EDM, this process has an added attraction in that the machined surface is stress-free and has high surface finish (0.5 micron). This characteristic is useful in the sinking of dies in hardened and tempered die blocks.

In the case of hard materials especially, ECM offers a higher rate of material removal as compared with conventional machining methods.

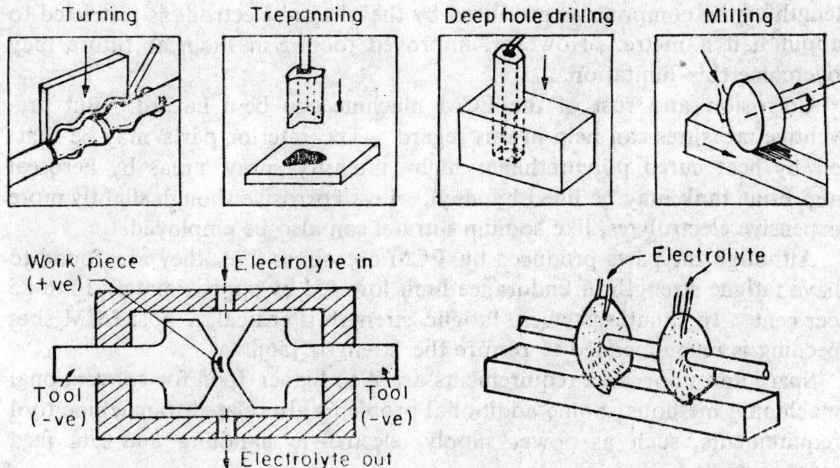

Fig. 3.9 Applications of ECM. The process is also being successfully used for machining gas and steam turbine blades

The fact that there is little or no wear of the tool in this process is advantageous because a large number of components can be machined without the tool having to be replaced.

Applications

Of the modern metal removal processes described in this book ECM is one of the most highly developed. It appears to have the greatest potential for versatility and is now being commercially utilized.

The process can handle a large variety of materials, limited only by their electrochemical properties and not their strength. Metal removal rates are high, especially of high strength, difficult-to-machine alloys developed for the needs of the space age. Fragile parts, which are otherwise not easily machinable, can be shaped by ECM.

The properties which make this process so widely applicable are (i) stress-free machining, (ii) burr-free machining, (iii) no tool wear, as there is no tool-work contact, and (iv) no burning or thermal damage to work piece.

Some specific applications (Fig. 3.9) of ECM are for (i) facing and turning complex three-dimensional surfaces, (ii) die sinking, particularly deep narrow slots and holes, (iii) profiling and any odd shape contouring, (iv) multiple hole drilling (v) trepanning, (vi) broaching, (vii) deburring, (viii) grinding, (ix) honing, and (x) cutting off.

Limitations

The main limitation of this process is that non-conductive materials cannot be machined. Another limitation is the inability to machine sharp interior edges and corners (less than 0.2 mm radius) because of very high current densities at those points. Blind holes cannot be machined in the solid block in one stage. Simple and straight-forward shapes can be rapidly produced by conventional machining methods rather than by ECM. The length of the components produced by the shaped electrode is restricted to about half a metre. However, improved tooling in the near future may overcome this limitation.

Corrosion and rust of the ECM machine can be a hazard. But preventive measures can help in this regard. The interior parts may be coated by heat cured polyurethane; high intensity spray areas by keroseal and brine tank may be lined by vinyl. Less corrosive though slightly more expensive electrolytes, like sodium nitrate, can also be employed.

Although the parts produced by ECM are stress-free, they are found to have fatigue strength or endurance limit lowered by approximately 10 to 25 per cent. In situations where fatigue strength is critical, a post-ECM shot peening is recommended to restore the strength [56].

Space and floor area requirements are also higher than for conventional machining methods. Some additional problems are related to machine tool requirements, such as power supply, electrolyte handling and tool feed servo systems.

Process Faults

Visual imperfections on the component surface are common process faults. Machines which are equipped with very sensitive spark detection and protective system can detect faults and shut the machine down without visible spark damage on the work or tool surface. In order to identify the cause of the fault by its position and appearance, it may then be necessary to slow down the machine switching time so that a large enough spark marks the work at the point of the process fault. Experienced operators can normally identify the cause without much difficulty. The main process faults are (i) cavitation, (ii) bright spots, (iii) poor work surface finish and (iv) work inaccuracy. Table 3.2 gives a check list of process faults and recommendations for corrective action.

Table 3.2

Check List of ECM Process Faults

S. No. (1)	Fault (2)	Location (3)	Cause (4)	Sequence of corrective action (5)
(i)	Machine stops operating. No apparent damage.		Small spark insufficient to damage tool.	Reduce sensitivity of protective circuit to increase damage to tool location of fault.
(ii)	Small irregular raised area, often with bright work surface. As above with	Sharp change in tool geometry or in sharply divergent flow path.	Cavitation.	Reduce electrolyte supply pressure. Blend out sharp radii on tool. Increase tool feed. Redesign tool for convergent electrolyte flow.

(Contd.)

Table 3.2 *(Contd.)*

(1)	*(2)*	*(3)*	*(4)*	*(5)*
	spark damage. Uniform patterned raised area (like a fossilized shell).			
(iii)	Striation, ripples on work surface.	Near fluid entry to tool, or in divergent flow areas.	Anode potential sensitivity to flow velocity variations.	Increase electrolyte supply pressure. Reduce tool feed rate. Reduce voltage to maintain the same gap. Select alternative electrolyte.
(iv)	Slightly poor surface finish.	Surfaces most closely angled to tool feed direction.	Differential machining of material phases.	Increase tool feed rate. Increase voltage to maintain gap size. Dilute electrolyte and increase voltage.
(v)	Poor surface finish.	All surfaces.	Differential machining of material phases.	Select alternative electrolyte.
(vi)	Consistent work inaccuracy.	All surfaces of work.	Incorrect tool alignment or tool contour error.	Reinspect tooling for accuracy of contour and alignment.
		Positions relating to original material shape.	Inadequate material allowance on part prior to ECM.	Allow correct material allowance.
		Inaccuracies on surfaces most closely angled to tool feed direction.	Incorrect operating gap or incorrect design of tool contour.	Check that design gap is operating. Check calculations used in designing tool contour.
		Inaccuracies increase with electrolyte flow path length.	Electrolyte conductivity variation across work surface.	Increase supply pressure of electrolyte. Decrease tool feed rate.
		At work boundaries or sharp work features.	Electrical field concentration of machining current.	Insulate all inoperative areas of tool. Add surplus material to work blank at field concentration areas.
		Isolated thin section of component.	Voltage loss in transmitting machining current through component.	Apply additional contacts near area of inaccuracy where applicable. Apply maximum available machining voltage and dilute electrolyte. Reduce tool feed rate

(Contd.)

Table 3.2 (*Contd.*)

(1)	(2)	(3)	(4)	(5)
				with corresponding electrolyte dilution to maintain gap size.
(vii)	Random inaccuracy in work.	Random areas.	Overheating of electrical contacts to work causing distortion.	Check for contact burning. Increase contact pressure.
		Occurring in similar areas.	Distortion of part due to electrolyte pressure or clamping forces, or residual stresses in part.	Check distortion as work is unclamped after machining. Stress relieve work before machining. Rearrange clamping. Provide additional support of work against electrolyte forces.
(viii)	Spark damage to tool or work.	Near point of electrolyte entry into machining gap.	Foreign particles in electrolyte.	Check internally filters, electrolyte ducts, seals and joints.
		At location of some other process fault.	Insulation failure within tool. Extreme condition of a previously considered fault.	Check insulated surfaces of tool. Identify and eliminate basic process fault.

ELECTROCHEMICAL GRINDING (ECG)

This process employs a grinding wheel in which an insulating abrasive is set in a conducting bonding material. The D.C. power is connected to the part and the conductive bond of the grinding wheel in such a way that the latter is at negative potential with respect to the component part. Brushes are used on the grinder spindle for the supply of current into the spindle, from which it then flows to the grinding wheel. The region between the wheel and work piece is flooded with electrolyte. The schematic diagram of the process is shown in Fig. 3.10.

When the work piece contacts the bed of the machine, an electrolytic cell is formed, with the work piece as anode and the body of the grinding wheel as cathode. The insulating abrasive particles in the grinding wheel protrude evenly above the wheel surface, and when the work piece is pressed into contact with these, the height of the abrasive particles above the wheel determines the effective gap between the anode and cathode. It is in this space that electrolysis actually takes place.

A D.C. voltage of about 5-15 V is applied between the work piece and the grinding wheel. Current densities range from 2 A/cm^2 in grinding tung-

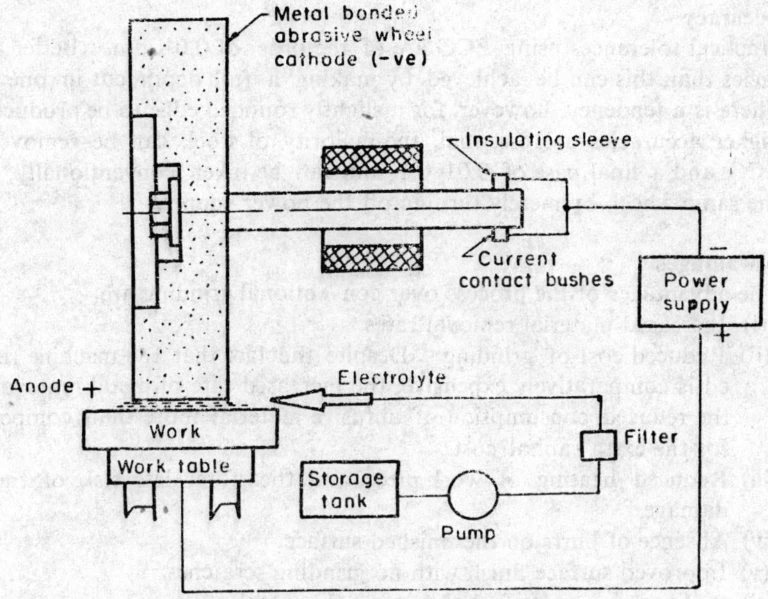

Fig. 3.10 Schematic of electrolytic grinding process

sten carbide to about $3 A/cm^2$ in grinding steels.

The electrolyte used in this process does not differ from that employed in electrochemical machining.

Material Removal

Most of the material removal in electrolytic grinding is by electrochemical action. Some of the metal is mechanically removed by the abrasive which is in contact with the work. The main functions of the abrasive particles are:

(i) To provide insulation between the anode and cathode and to determine the effective gap thickness between them.

(ii) To continually remove any passive layer that may be formed on the work piece.

In general, practical metal removal rates with ECG are of the order of $1.0 \ cm^3/min/100A$. However, $0.5 \ cm^3/min/100 \ A$ is often used for approximation.

Surface Finish

Surface finish produced by ECG on tungsten carbides can be expected to range from 0.2-0.4 microns for plunge grinding, and 0.4-0.5 microns for surface or traverse grinding. In the case of steels and various alloys, surface finish up to 0.4-0.6 microns can be easily obtained. Generally speaking, the higher the hardness of alloy, the better is the finish.

Accuracy

Practical tolerances using ECG are of the order of 0.01 mm. Better accuracies than this can be achieved by making a full depth cut in one pass. There is a tendency, however, for a slightly rounded edge to be produced. If higher accuracies are essential, the majority of stock can be removed by ECG and a final pass of 0.01–0.1 mm can be taken conventionally, with the same wheel, by merely turning off the power supply.

Advantages

The advantages of the process over conventional grinding are:

 (i) Increased material removal rates.
 (ii) Reduced cost of grinding. Despite the fact that the machine required is comparatively expensive, the increased rate of metal removal and the reduced consumption of abrasive material more than compensate for the extra capital cost.
(iii) Reduced heating of work piece, and therefore, less risk of thermal damage.
(iv) Absence of burrs on the finished surface.
 (v) Improved surface finish with no grinding scratches.
(vi) Reduced pressure of work against the wheel.

Applications

(i) *Carbide Cutting Tools*

The process of ECG is most extensively used for the grinding of carbide cutting tools. In comparison with conventional grinding, ECG provides savings of about 75 per cent in wheel costs and about 50 per cent in labour costs in grinding tungsten carbide.

(ii) *Fragile or Very Hard and Tough Materials*

ECG is particularly useful for grinding fragile parts such as honeycomb, thin walled tubes and skins, hypodermic needles, etc. Further, high production rates can be achieved when grinding hard, tough, stringy, work-hardenable or heat-sensitive materials.

ELECTROCHEMICAL DEBURRING

In almost all forming and machining operations very fine burrs of metal are invariably left on the work piece. These burrs are undesirable, particularly for precision components, as they may break loose and disturb a delicately balanced mechanism. They are also dangerous for the fingers. These burrs have been successfully removed through electrochemical means by a process called 'electrochemical deburring'.

Process

An electrode is positioned close to the area of the work piece (made of conducting material) to be deburred. The electrode is connected to the negative, and the work piece to the positive terminal of a D.C. source. An electrolyte is made to flow between the electrode and the work. Burrs are removed by electrochemical action.

Application

The process is useful in many precision engineering industries. For example, valve electrodes in American industries are being deburred by this process. Automatic machines are now available which can handle as many as 500-1000 parts per hour, depending upon the components size, position and amount of burrs to be removed. The process is most applicable to mass production industries where production runs are long so that extra costs incurred on special tooling required for the process are amortized over a large number of parts.

In specific applications, this process has drastically reduced the operation time. The cost has fallen by more than 50 per cent but more significantly, the process ensures complete burr removal in contrast with random deburring that is obtained by manual methods. The process is most advantageous for those applications where burrs cannot be easily removed manually.

ELECTROCHEMICAL HONING

Process

In this process, the stock removal capabilities of ECM are combined with the accuracy capabilities of honing. The basic principles of the process regarding voltage, current, electrolyte and materials processed are the same as those described under ECM. The process, as shown in Fig. 3.11, consists in rotating and reciprocating the tool inside the cylindrical components. The electrolyte is fed under pressure through holes in the tool so that at every point there is uniform flow and velocity. The gap between the tool and the work piece is usually adjusted by the use of an expanding tool. At

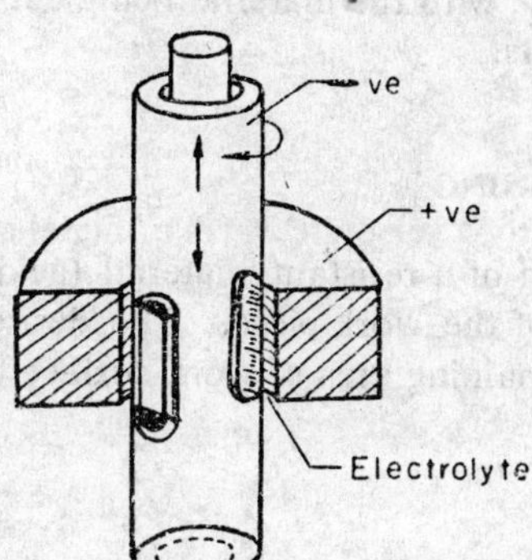

Fig. 3.11 Schematic of electrolytic honing process

the start of the operation, the gap is approximately 1 mm and increases during the cycle.

Bonded abrasive honing stones are forced out with equal pressure in all directions from slots in the tool. These stones are non-conductive and assist in electrochemical action. They also abrade the residue left by the electrochemical action. A clean surface thus generated is ready for further chemical attack. If the work piece is not round but tapered or wavy, the stones cut away all high areas and remove the geometric error.

As in conventional honing, the abrasive should be selfdressing. If the stones glaze, the stock removal rate is reduced. If they dress too rapidly, the operating cost becomes high. The design of the tool and the adjusting head mechanism is important for proper abrasive application.

Accuracy and Surface Finish

Size tolerance of 0.01 mm on the diameter can be obtained and roundness can be maintained at less than 0.005 mm. Surface roughness of the order of 0.1-0.5 micron CLA is obtainable by this method but if it is required to have a specified roughness, the stones are put to work for a few seconds after the power is switched off. The surface finish obtained in this manner is dependent upon the size of abrasive grains, speed of rotation and reciprocation, and duration of the 'run out' period.

Advantages

The advantages of electrochemical honing are similar to those claimed for electrolytic grinding: increased metal removal rate particularly on hard materials, burr-free action, less pressure required between stones and work, reduced noise and distortion when honing thin-walled tubes, cooler action leading to increased accuracy with less material damage.

Applications

The process is easily adaptable to cylindrical parts for trueing the inside surfaces. The size of the cylinder that can be processed by this method is limited only by the current and electrolyte that can be supplied and distributed. Any surface roughness compatible with the material being cut is duplicated over a number of component parts.

CHEMICAL MACHINING

Chemical machining involves the application of a resistant material (acidic or alkaline in nature) to certain portions of the work piece. The desired amount of material is removed from the remaining area of work surface by the subsequent application of an etchant.

Elements of Process

Basically, there are only two elements of the process:

    (i) resists or maskants, (ii) etchants.

Resists (Maskants)

There are three types of resists: cut and peel, photographic and screen resists.

The cut and peel resists are first applied to the entire part by the spray or dip method. It is then cut from the areas to be etched. Due to the inherent nature of the maskant and thickness of the coating applied (up to 0.2 mm) high chemical resistance is obtained permitting etching depths up to 1.5 mm. This type of maskant is not used in applications where critical dimensional tolerances are required. The maskants currently available include vinyl, neoprene and butyl base materials.

Photographic resists produce etchant resistant images by means of photographic techniques. When exposed through a high contrast negative, the materials can generally produce a positive or negative image of the negative itself. These resists are usually applied in liquid form by the spray, dip or roll coating techniques. These resists are used for:

(i) Thin materials up to 0.8 mm thick.

(ii) Parts requiring dimensional tolerances of the etchant resistant image tighter than $\pm$ 0.1 mm and up to 1×1.5 m section.

(iii) Automatic processing of high volume components.

Screen resists are materials which can be used on the workpiece through normal silk screening techniques. The image accuracies are better than can generally be achieved by other types of maskants. Screen printing, is generally limited to:

(a) Parts not larger than 1×1 m.

(b) Parts having only flat surfaces or simple moderate contours.

(c) Parts where depth of etch does not exceed 1.5 mm in depth from one side.

(d) Parts which do not require etchant resistant image accuracy greater than $\pm$ 0.2 mm.

The selection of a resist for use in chemical machining depends on the following factors:

(i) Chemical resistance required.

(ii) Number of parts to be produced.

(iii) Detail or resolution required.

(iv) Shape and size of component.

(v) Ease of removal.

(vi) Economics.

Etchant

The basic function of an etchant is to convert a material (say a metal) into a metallic salt that can be dissolved in the etchant, and thus removed from the work surface. Table 3.3 gives some commonly used etchants, their characteristics, and the materials for which they are used.

The factors which affect the selection of an etchant for a given component are:

(i) Material to be etched.
(ii) Type of maskant or resist used.
(iii) Depth of etch.
(iv) Surface finish required.
(v) Potential damage to metallurgical properties of work.
(vi) Rate of material removal.
(vii) Economics.

Table 3.3

Etchant Characteristics and General Applications

Etchant	Concentration	Temp. (K)	Etch rate (cm/min) fresh solution	Etch factor	Metals that etchant will attack
$FeCl_3$	12 to 18° Bé*	320°	.002	1.5:1 to 2.0:1	Aluminium alloys
HCl, HNO_3H_2O	10:1:9	320°	.002 to .004	2:1 (variable)	
$FeCl_3$	42°Bé	320°	.002	2:1	Cold rolled steels
HNO_3	10 to 15% (vol.)	320°	.002	1.5 to 2.0:1	
$FeCl_3$	42°Bé	320°	.004	2.5 to 3.0:1	Copper and its alloys
$(NH_4)_2S_2O_8$	0.22 g/cm³ H_2O	start at 300-320°	.002	2 to 3:1	
Chromic acid	Commercially available	325°	.0030	2 to 3:1	
$CuCl_2$	35° Bé (regenerated)	325°	.001	2.5 to 3:1	
HNO_3	12 to 15% (vol.)	300°-320°	.002 to .004	—	Magnesium
$FeCl_3$	42°Bé	320°	.001 to .002	1:1 to 3:1	Nickel
$FeCl_3$	42°Bé	325°	.002	1.5 to 2:1	Stainless steel, tin
HNO_3	10 to 15% (vol.)	320 to 325°	.002	—	Zinc

*Baumé specific gravity scale.

Advantages

The advantages of this process are:

(i) Several parts can be machined simultaneously.
(ii) Machined or extruded parts may be reduced on selected areas or all over by ChM to provide web sections that are thinner than those obtainable by conventional methods.
(iii) Castings may be designed uniformly oversized, heat treated with little or no warpage, and then chemically machined to achieve final desired dimensions.
(iv) Extrusions, forgings, castings, formed sections, and deep drawn parts can be lightened considerably by ChM.

Applications

Chemical machining has been applied successfully in many cases where the depth of material removal is critical to a few microns and the tolerances are close. The surface finish obtained in the process is of the order of 0.5-2 microns. The process is usefully applied to:

(i) Remove metal from a portion or the entire surface of formed or irregularly shaped parts, such as forgings, castings, extrusions, or formed wrought stock.

(ii) Reduce web thickness below practical machining, forging, casting or forming limits.

(iii) Taper sheets and pre-formed shapes.

(iv) Produce stepped webs.

(v) Engraving on any metal piece.

REVIEW QUESTIONS

1. What is the principle of electrochemical machining? What are the materials commonly used for making a tool for use in this method? Is there any limitation on the type of material that can be machined by ECM?
2. What are the functions of an electrolyte? What factors need to be considered while selecting it? Discuss the advantages and limitations of some electrolytes.
3. (i) Describe the chemistry involved in the ECM process.
   (ii) Derive a theoretical relationship for the determination of the metal removal rate in ECM.
   (iii) What is the 'self adjusting feature' in ECM?
4. Derive an equation for the maximum permissible feed rate of the cathode tool, and hence, deduce the relation for the electrolyte temperature change for a given feed rate of cathode tool.
5. Write short notes on:
   (i) The economics of electrochemical machining.
   (ii) The effect of high temperature and pressure of electrolyte in the ECM process.
   (iii) Applications of electrolytic grinding process.
6. (i) What are the factors on which the selection of a resist for use in chemical machining depend?
   (ii) Distinguish between cut and peel resists and photographic resists.
7. What are the specific advantages of using chemical machining over electrochemical machining? Give some practical applications of the chemical machining process.

4

Thermal Metal Removal Processes

Several machining processes involving the application of very intense local heat have come into use in recent years. In these processes, material is removed by melting or vapourizing small areas at the surface of the work-piece.

The processes in which metal removal is based on thermal principles are:
   (i) Electric Discharge Machining (EDM).
   (ii) Plasma Arc Machining (PAM).
   (iii) Electron Beam Machining (EBM).
   (iv) Laser Beam Machining (LBM).
   (v) Hot Machining.

The principle of working, applications and essential characteristics of these processes have been discussed in this chapter.

ELECTRIC DISCHARGE MACHINING (EDM)

In 1970, the English scientist, Priestley, first detected the erosive effect of electrical discharges on metals. More recently, during research (to eliminate erosive effects on electrical contacts) the soviet scientists, Lazarenko and Lazarenko, decided to exploit the destructive effect of an electrical discharge and develop a controlled method of metal machining. In 1943, they announced the construction of the first spark erosion machine. The spark generator used in 1943, known as the Lazarenko circuit, has been employed over many years in power supplies for EDM machines and an improved form is being used in many current applications.

The EDM process can be compared with the conventional cutting process, except that in this case, a suitably shaped tool electrode, with a precision controlled feed movement is employed in place of the cutting tool, and the cutting energy is provided by means of short duration electrical impulses. EDM has found ready application in the machining of hard metals or alloys which cannot be machined easily by conventional methods. It thus plays a major role in the machining of dies, tools, etc., made of tungsten carbides, stellites or hard steels. Alloys used in the aeronautics industry, for example, hastalloy, nimonic, etc., could also be machined conveniently by this process. This process has the added advantage of being capable of machining complicated components.

The popularity of the EDM process is due to the following advantages:

(i) The process can be readily applied to electrically conductive materials. Physical and metallurgical properties of the work material, such as strength, toughness, microstructure, etc., are no barrier to its application.

(ii) During machining, the work piece is not subjected to mechanical deformation as there is no physical contact between the tool and work. This makes the process more versatile. As a result, slender and fragile jobs can be machined conveniently.

(iii) Although the metal removal in this case is due to thermal effects, there is no heating in the bulk of the material.

(iv) Complicated die contours in hard materials can be produced to a high degree of accuracy and surface finish.

(v) The overall production rate compares well with the conventional processes because it can dispense with operations like grinding, etc.

(vi) The surface produced by EDM consists of a multitude of small craters. This may help in oil retention and better lubrication, specially for components where lubrication is a problem. The random distribution of the craters does not result in an appreciable reduction in fatigue strength of the components machined by EDM.

(vii) The process can be automated easily requiring very little attention from the machine operator.

Spark Erosion Machining Processes

Electric Discharge Machining (EDM) is the removal of materials conducting electricity by electrical discharges between two electrodes (workpiece electrode and tool electrode), a dielectric fluid being used in the process. The aim of the process is controlled removal of material from the work piece.

Figure 4.1 shows a classification of the spark erosion machining processes.

Sinking by EDM

In this case, the metal removal is affected by nonstationary electrical discharges which are separated from each other both spatially and temporarily. This process includes those EDM operations in which the average relative

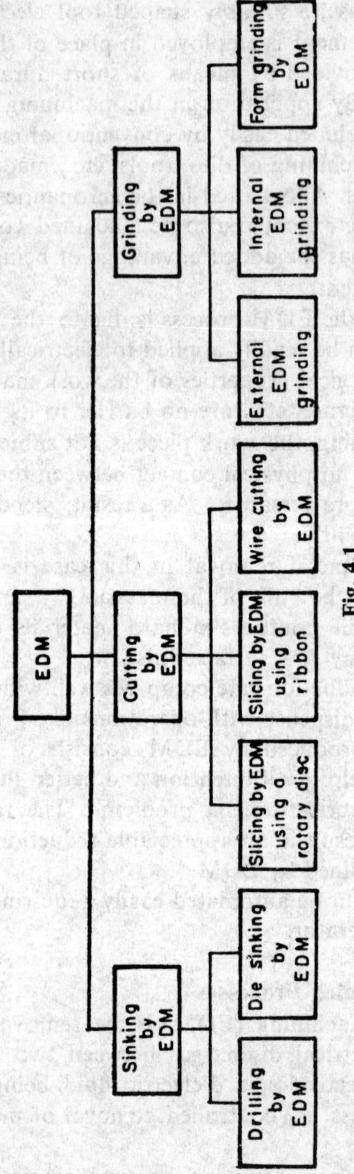

EDM
Sinking by EDM
Cutting by EDM
Grinding by EDM
Drilling by EDM
Die sinking by EDM
Slicing by EDM using a rotary disc
Slicing by EDM using a ribbon
Wire cutting by EDM
External EDM grinding
Internal EDM grinding
Form grinding by EDM

Fig. 4.1

speed between the tool and work piece is coincident with the penetration
speed in the workpiece (Fig. 4.2).

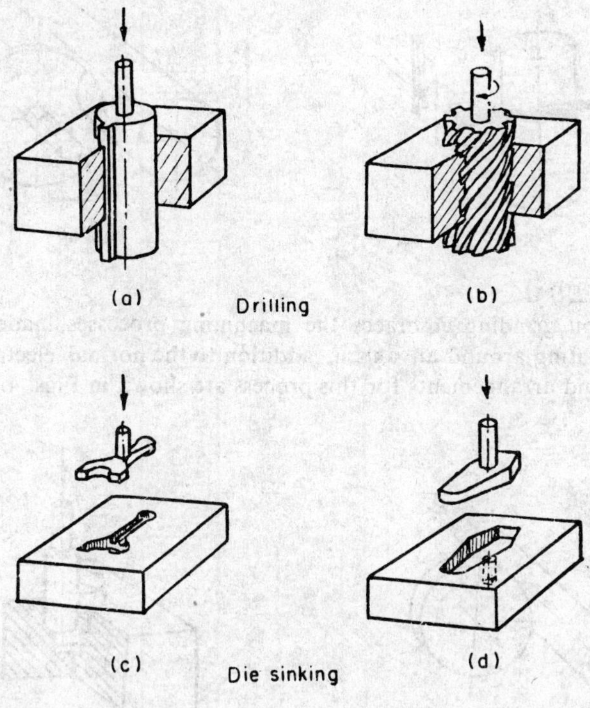

Fig. 4.2

Cutting by EDM

It includes those machining operations where the workpiece is cut off or
notched. Figures 4.3 to 4.6 demonstrate various EDM cutting operations.

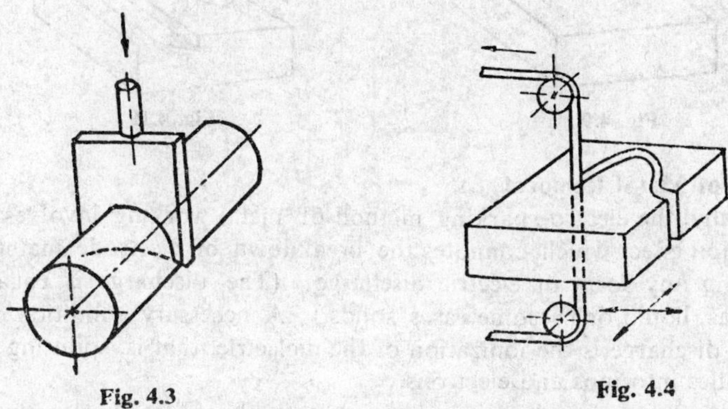

Fig. 4.3

Fig. 4.4

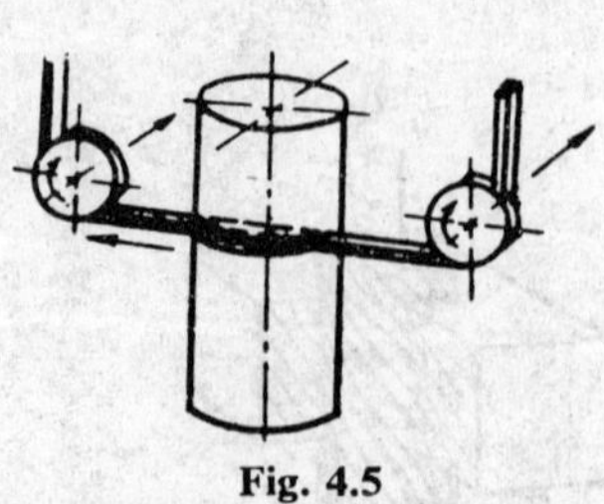

Fig. 4.5

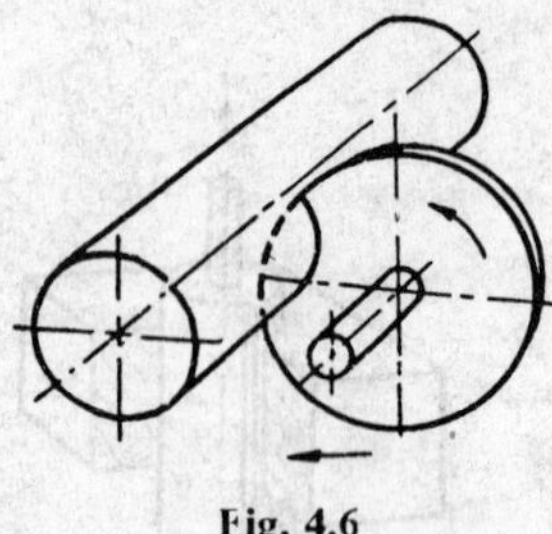

Fig. 4.6

Grinding by EDM

Spark erosion grinding embraces the machining processes made with an electrode rotating around an axis in addition to the normal electrode feed. The forms and arrangements for this process are shown in Figs. 4.7 to 4.10.

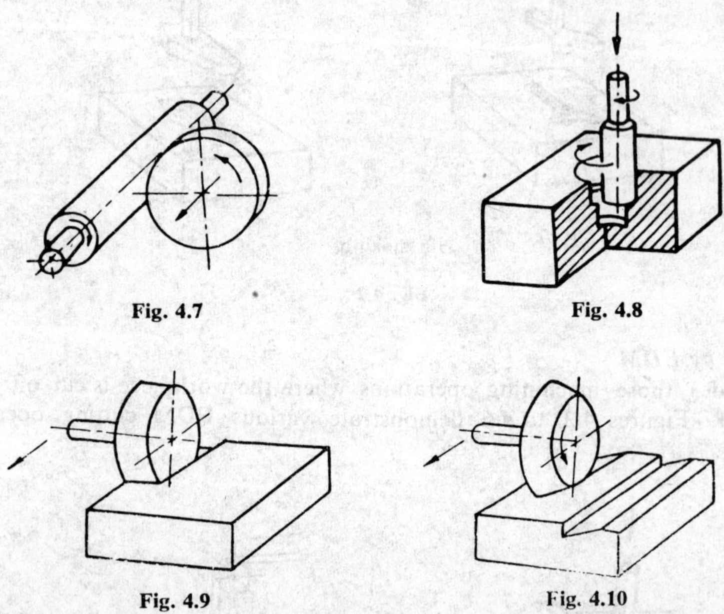

Fig. 4.7

Fig. 4.8

Fig. 4.9

Fig. 4.10

Mechanism of Metal Removal

Fundamentally, the electro-sparking method of metal working involves an electric erosion effect which connotes the breakdown of electrode material accompanying any form of electric discharge. (The discharge is usually through a gas, liquid or in some cases solids.) A necessary condition for producing a discharge is the ionization of the dielectric, that is, spilitting up of its molecules into ions and electrons.

Consider the case of a discharge between two electrodes (tool cathode

and work anode) through a gaseous (Fig. 4.11) or liquid medium. As soon
as suitable voltage is applied across the electrodes, the potential intensity of
the electric field between them builds up, until at some predetermined value,
the individual electrons break loose from the surface of the cathode and
are impelled towards the anode under the influence of field forces (Fig. 4.11).
While moving in the inter-electrode space, the electrons collide with the
neutral molecules of the dielectric, detaching electrons from them and causing ionization. At some time or the other, the ionization becomes such
that a narrow channel of continuous conductivity is formed. When this
happens, there is a considerable flow of electrons along the channel to the
anode, resulting in a momentary current impulse or discharge. The liberation
of energy accompanying the discharge leads to the generation of extremely
high temperature, between 8,000° and 12,000°C, causing fusion or partial
vapourization of the metal and the dielectric fluid at the point of discharge.
The metal in the form of liquid drops is dispersed into the space surrounding the electrodes by the explosive pressure of the gaseous products in
the discharge. This results in the formation of a tiny crater at the point of
discharge in the workpiece.

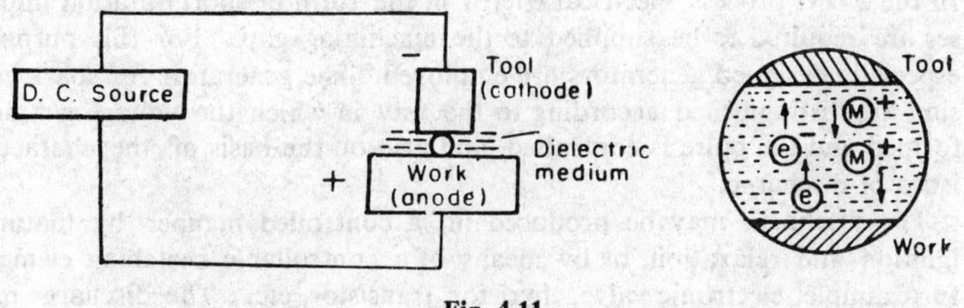

Fig. 4.11

Comparatively less metal is eroded from the cathode (tool) as compared
to the anode work due to the following reasons:

(i) The momentum with which positive ions strike the cathode surface is
much less than the momentum with which the electron stream impinges on
the anode surface.

(ii) A compressive force is generated on the cathode surface by the spark
which helps reduce tool wear.

Most of the EDM operations are conducted with electrodes (tool and
work) immersed in a liquid dielectric, for example paraffin, and the mechanism of sparking is similar to that described above except that the dielectric
is contaminated with conductive particles. Furthermore, the particles removed from the electrodes due to the discharge fall in the liquid, cool down
and contaminate the area around the electrodes by forming colloidal suspensions of metal. These suspensions, along with the products of decomposition of the liquid dielectric are drawn into the space between the
electrodes during the initial part of the discharge process and are distributed

along the electric lines of force, thus forming current carrying 'bridges'. Discharge then occurs along one of these bridges as a result of ionization, described earlier.

Spark discharge in liquid leads to an intense ejection of anode particles into the surrounding space, but discharge in a gas results in the partial transfer and diffusion of detached anode particles into the surface of the cathode. Both these phenomena are used in metal working; the first in performing dimensional working operations, for example, drilling, die sinking and the preparation of tool, etc.; the second is employed in operations connected with the toughening and building up of surfaces. The spark erosion process must be visualized as a succession of spark discharges distributed over the surface to be eroded. The spark will pass between the electrode and work piece at that particular point at which the electric field strength in the inter space is highest. Thus, successive spark discharges erode the entire surface. A surface produced by this method has a pitted appearance, the size and depth of the pits are determined by the spark energy, the nature of workpiece material and the dielectric.

Spark Erosion Generators

In the EDM process, electrical energy in the form of short duration impulses are required to be supplied to the machining gap. For this purpose, especially designed generators are employed. The generators for spark erosion are distinguished according to the way in which the voltage is transformed and the pulse is controlled, and also on the basis of the characteristics of discharge.

The discharge may be produced in a controlled manner by 'natural' ignition and relaxation, or by means of a controllable switching element, for example, electronic valve, thyristor, transistor, etc. The discharge may take place with constant or changing polarity.

On the basis of these facts, generators for EDM can be classified into:

(i) Relaxation generators.

(ii) Rotary pulse generators.

(iii) Static pulse generators.

Relaxation Generators

The relaxation or the R-C circuit was the first to be used in EDM. The circuit (Fig. 4.12) comprises a D.C. power source that charges a capacitor, 'C' across a resistance 'R'. If the condenser is initially uncharged and the D.C. supply is switched on, a heavy current will flow into the circuit with the condenser voltage rising continuously, as shown in Fig. 4.13.

The condenser voltage at the instant t can be described by the relationship

$$U_{1(t)} = U_{B}[1 - e^{-t/RC}] \qquad (4.1)$$

Equation 4.1 predicts that the condenser voltage will approach the supply voltage (U_s) with a time constant equal to RC and after $t = RC$, the condenser voltage will be 63 per cent of the supply voltage (U_s). A discharge, across the working gap will occur if $U_{1(t)}$ equals the breakdown voltage (U_b) of the dielectric within the gap. After the discharge, the dielectric deionizes, the capacitor is recharged and the cycle repeats itself. The time taken to recharge the capacitor to the breakdown voltage must be sufficient to allow the dielectric to deionize.

In practice, the spark gap is adjusted so that the discharge takes place corresponding to a gap voltage of about 0.72 U_s. Although a higher gap voltage would liberate much more energy, the time required to recharge the condenser increases. Thus, the benefit from a higher energy content per spark is more than offset by a reduction in the number of condenser discharge per unit time. It has been found in an R-C circuit, that for a given condenser and breakdown voltage, there exists a certain value of 'R' that will ensure correct length of the charging cycle.

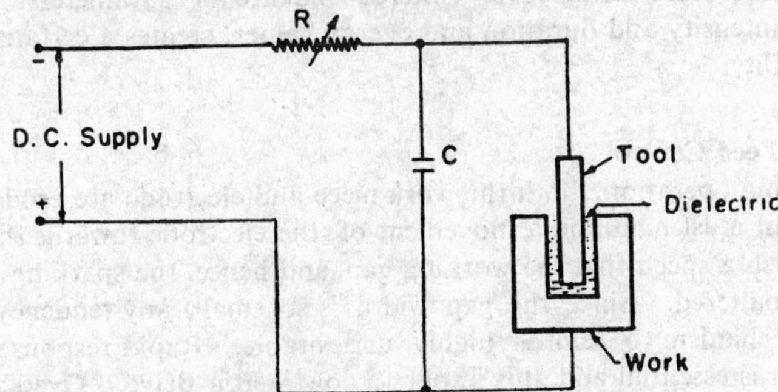

Fig. 4.12 Elementary relaxation circuit for EDM

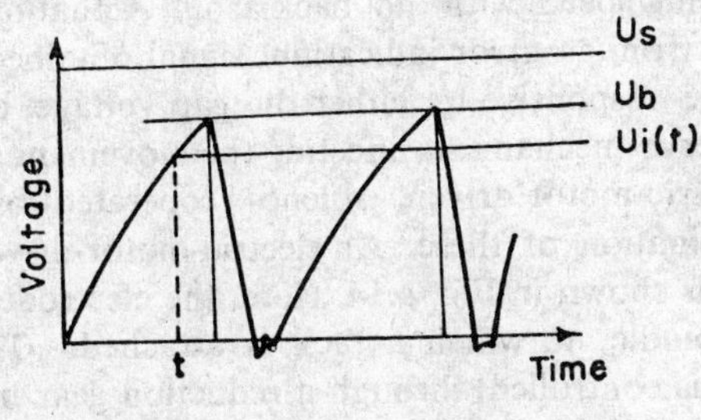

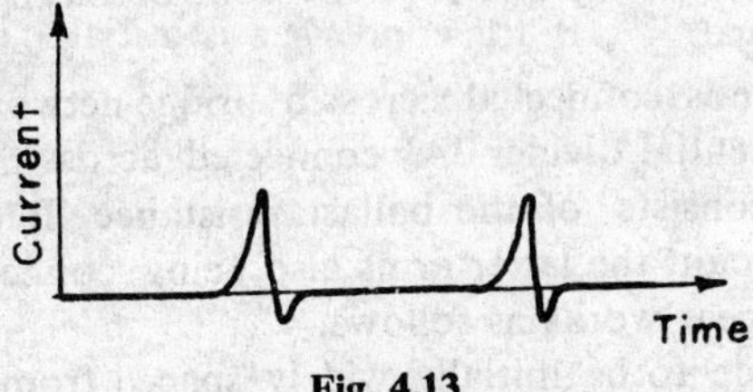

Fig. 4.13

In a relaxation generator, the spark repetition rate, for a given supply voltage and capacitance, cannot be increased beyond a critical value and is determined by the speed at which the spark gap is deionized and cleared of the debris after each discharge. Forced circulation of the dielectric through the gap is necessary if high metal removal rates are desired. As the working gap is of the order of 0.025-0.050 mm, forced circulation is difficult, especially when large electrodes are involved. In such cases, a lower erosion rate must be accepted than is possible with small size electrodes.

The fundamental advantages of relaxation circuits are their comparative cheapness, simplicity of design, robustness and relatively extensive range of discharge. They remain the only practical means of generating low energy ranges and high frequencies required for fine finishing and delicate operations.

In spite of many modifications of relaxation circuits, they are liable to result in high tool wear and slow metal removal rates, compared with other types of generators. Moreover, interdependence of parameters, such as discharge intensity and duration and energy values, creates a certain degree of inflexibility.

Electrode Feed Control

Since, during operation, both the work piece and electrode are eroded, the feed control must maintain a movement of the electrode towards the work piece at such a speed that the working gap, and hence, the sparking voltage remains unaltered. Since the gap width is so small, any tendency of the control mechanism to hunt is highly undesirable. Rapid response of the mechanism is essential and this implies a low inertia drive. Overshooting may completely close the gap and cause a short circuit; hence, it is essential to have rapid reversing speed with no backlash. Actuation of the control drive is derived from an error indication signal obtained from an electrical sensing device responsive to either the gap voltage or the working current or both. Servo-mechanisms affecting the movement of the electrode may be either electric-motor-driven, solenoid operated or hydraulically operated or a combination of these. An electric-motor-driven type of gap control mechanism is shown in Fig. 4.14. Here, the electrode is carried in a chuck fixed to a spindle, to which a rack is attached. The axial movement of the spindle is controlled through a reduction gear box driven by a D.C. shunt motor, which is reversible so that the electrode can be withdrawn, should the gap be bridged by swarf or the control mechanism cause the electrode to overshoot.

The motor armature is connected across a bridge network, the arms of which consist of a potential divider 'A' connected across the D.C. supply, while the other arm consists of the ballast resistance 'B' and condenser 'C' of the charging circuit, the latter arms also being connected across the supply. The control gear works as follows.

Assume the electrode to be initially widely spaced from the work piece

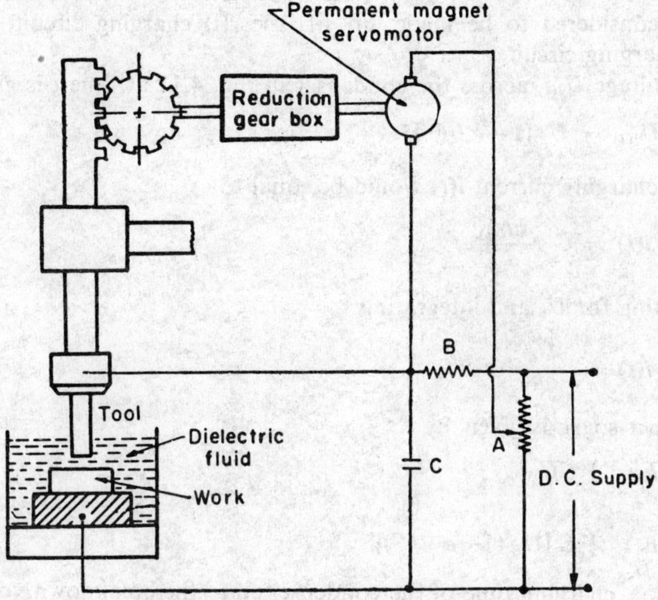

Fig. 4.14   Electrode feed control in EDM

and the current supply switched on to the condenser. This will cause the condenser to be charged and the voltage will rise to approach the supply voltage. The supply voltage will, therefore, prevail across one lower arm of the bridge. The voltage across the other arm of the bridge will depend on the potentiometer setting, and if this setting is midway, then the voltage across the bridge (i.e. the difference between voltages across the two lower limbs) will be half the supply voltage. This voltage tends to rotate the motor, causing the electrode to close the gap. When the electrode reaches the correct position, sparking takes place and the condenser rapidly charges and discharges so that a saw-tooth wave-form is produced across its terminals. The electrode will cease to move when the average value of this voltage equals that prevailing across the lower limb of the potentiometer. Under this condition, the bridge is balanced and there is no armature current. Should the electrode overshoot, the gap width will be smaller and the average condenser voltage will fall since the condenser will no longer be able to charge up to the specific voltage. The bridge is now unbalanced with a reverse polarity so the motor reverses and widens the gap until the correct position is attained. If the electrode touches the work, the condenser is short circuited, causing the supply voltage to appear across the ballast resistance, and the electrode is lifted away from the work piece. A simiiar action takes place when the gap is bridged by swarf. The required gap width can be obtained, for a given operation, by adjusting the potentiometer setting.

Power delivered by an R-C circuit   The relaxation circuit shown in Fig. 4.12 can be considered to be made up of the (i) charging circuit, and the (ii) discharging circuit.

The voltage $U_{i(t)}$ across the condenser in Fig. 4.12 at time t is given by

$$U_{i(t)} = U_s [1 - e^{-t/RC}] \tag{4.1}$$

and the charging current $i(t)$ would be equal to

$$i(t) = C \, \frac{du_{i(t)}}{dt} \tag{4.2}$$

Substituting for U_s and integrating

$$i(t) = \frac{U_s}{R} \, e^{-t/RC} \tag{4.3}$$

Energy per spark is given by

$$E = \tfrac{1}{2} CU_b^2 \tag{4.4}$$

or

$$E = \tfrac{1}{2} C [U_s (1 - e^{-t/RC})]^2 \tag{4.5}$$

where $t' = $ charging time of the condenser up to the breakdown voltage. The power delivered in time t' (average) would be obtained as

$$E/t' = W_{avg} = \frac{C}{2t'} \left[U_s \left\{ 1 - e^{-t'/RC} \right\} \right]^2$$

For maximum power delivery through the circuit

$$\frac{dW_{avg}}{dx} = 0 \tag{4.6}$$

where

$$x = t'/RC$$

For maximum power, from Eq. 4.6, it is seen that $x = 1.26$
Substituting for x, we get

$$U_b = U_s [1 - e^{-1.26}]$$

or

$$\frac{U_b}{U_s} = 0.72 \tag{4.7}$$

Thus, it is seen that for maximum power delivery through the gap, the breakdown and supply voltage should follow the relationship given in Eq. 4.7.

Metal removal rate using relaxation circuit   Metal removal rate in EDM, using relaxation type circuit, is proportional to the product of frequency of charging (f) and the energy delivered per spark.

Metal removal rate $\propto f\,\tfrac{1}{2}\,CU_b^2$

or, Metal removal rate $= K_1\,f\,(\tfrac{1}{2}\,CU_b^2)$

where K_1 is the constant of proportionality.

The frequency of charging would be given by

$$f = 1/t'$$

where t', the time of charging the condenser is given by

$$t' = RC\,\log_e\left[\frac{1}{1-U_b/U_s}\right]$$

The metal removal rate (MRR) is, therefore, given by

$$\text{MRR} = \frac{K_1}{2R}\,U_b^2 \times \left[\frac{1}{\log_e\dfrac{1}{1-U_b/U_s}}\right] \tag{4.8}$$

From Eq. 4.8 and Fig. 4.15, it can be seen that for a given circuit, the metal removal rate will increase with decreasing R. However, R can not be made very low because, in that case, arcing will occur instead of sparking and such a situation is detrimental to the work surface finish. The minimum value of the resistance that will prevent arcing is known as critical resistance.

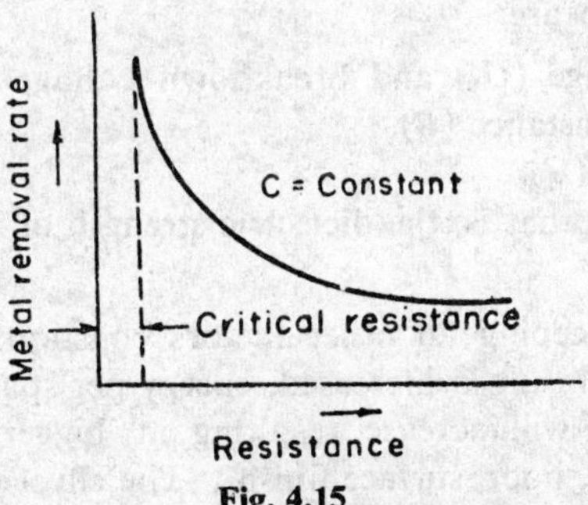

Fig. 4.15

Critical resistance If the equivalent inductance of the discharging circuit is considered to be L, then from the energy balance it is seen that

$$\tfrac{1}{2}\,Li^2_{d\,max} = \tfrac{1}{2}CU_b^2 \tag{4.9}$$

where $i_{d\,max}$ = Maximum discharge current $= \dfrac{U_c}{\sqrt{L/C}}$, and

$$U_c = \text{Condenser voltage at the instant of spark initiation} \tag{4.10}$$

But

$$i_{d\,max} = U_s - U_c/R$$

For a purely inductive circuit

$$i_{d\ max} = U_s/R_{min} \tag{4.11}$$

Equating Eqs. 4.10 and 4.11, we get the minimum value of the resistance (R_{min}) for a purely inductive circuit as

$$R_{min} = \frac{U_s}{U_c}\ \sqrt{L/C} \tag{4.12}$$

For

$$\frac{U_s}{U_c} = 1,\ R_{min} = \sqrt{L/C} \tag{4.13}$$

Equation 4.13 is found to be discordant with the experimental values. In practice, the circuit is not purely inductive, and hence, this equation should be used with modifications. The empirical relation given by Eq. 4.14 has been found to yield better results

$$R_{min} \geqslant 30\ \sqrt{L/C} \tag{4.14}$$

Electrical parameters in R-C circuit The theoretical analysis of the relaxation circuit given here indicates that the four parameters that govern the metal removal rate are:

(i) Supply voltage (U_s) and breakdown voltage (U_b).
(ii) Charging resistance (R).
(iii) Capacitance (C).
(iv) Gap setting, that is, the dielectric strength of the gap.

Supply voltage Keeping all other factors constant, an increase in breakdown voltage will result in increased, energy per spark. Consequently, the metal removal rate will increase, resulting in bigger craters on the work surface, and hence, poor surface finish. The effect of breakdown voltage and mean current on the metal removal rate is given in Fig. 4.16.

The selection of supply voltage is a compromise between several factors, for example, size of equipment, safety of operation, etc. The D.C. supply voltage used in EDM machines ranges between 30 and 200 V. Apart from the voltage and current, the electrode area for a given setting has been found to influence the rate of metal removal (Fig. 4.17). Figure 4.18 shows the effect of current density on the removal rate of steel when machined by brass tool.

Figure 4.19 shows the effect of pulse energy on the metal romoval rate when steel is machined by brass electrodes in kerosene medium. The pulse energy at a constant voltage is varied by changing the size of the capacitors used.

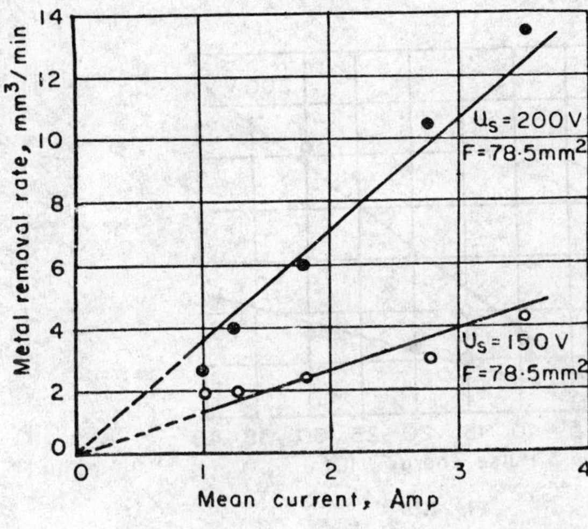

Fig. 4.16 Effect of voltage and current on the rate of metal removal. Dielectric: Kerosene; Tool: Brass; Work material: Low C steel; F: Tool electrode area.

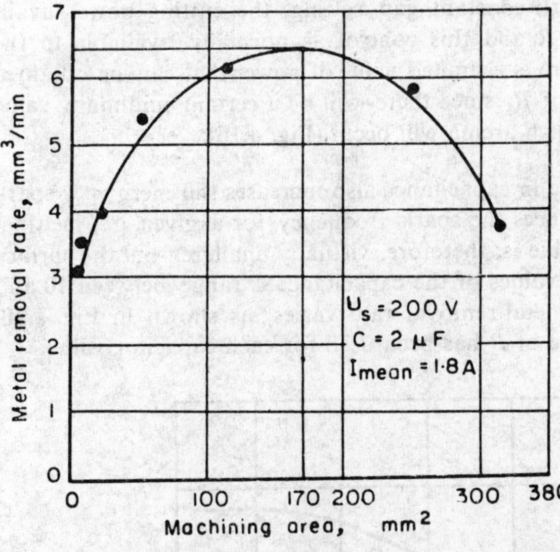

Fig. 4.17 Effect of machining area on the rate of metal removal. Work material: Low C steel; Tool: Brass; Dielectric: Kerosene.

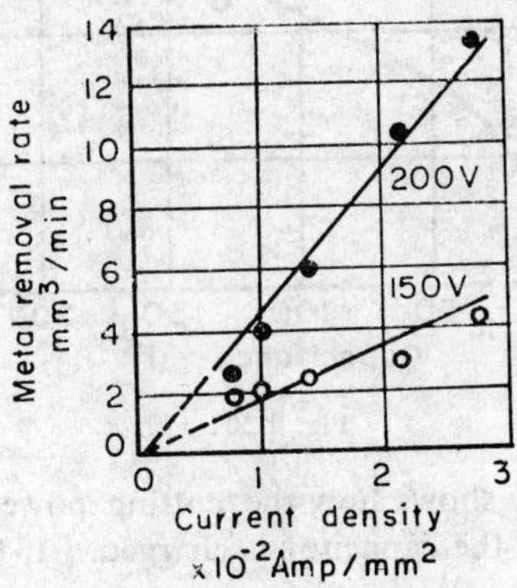

Fig. 4.18

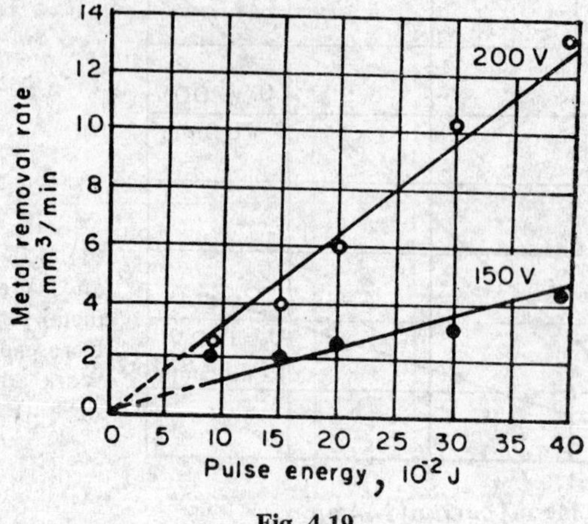

Fig. 4.19

Charging resistance With constant gap setting, the cutting power available varies inversely as R and this control is normally available to the operator. However, there is a limited value of power that can be obtained by decreasing the value of R, since there will be a certain minimum value of R ($= R_{min}$) below which arcing will occur (Fig. 4.15).

Capacitance An increase in capacitance also increases the energy per spark and at the same time reduces the spark frequency for a given gap setting. The cutting power available is, therefore, virtually unaltered but the surface finish deteriorates. The values of the capacitance C range between 10 and 100 microfarads. The metal removal rate varies, as shown in Fig. 4.20, where the minimum value of R has been used for each capacitor value.

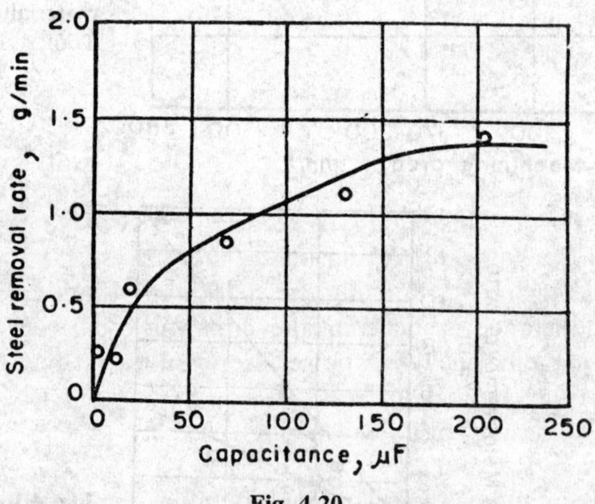

Fig. 4.20

Gap setting Figure 4.21 shows how the cutting power depends upon the relative voltage to which the capacitor is charged. If the gap is larger than

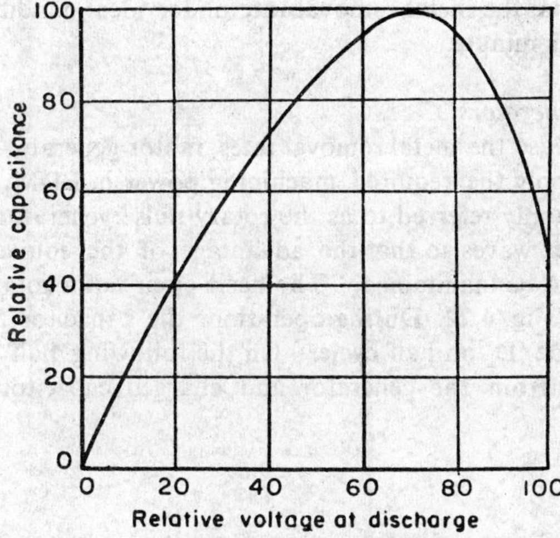

Fig. 4.21

the optimum, the consequent reduction in frequency is not compensated for by an increase in energy per spark and hence, the power falls. At shorter gaps the power decreases because the increased frequency is not sufficient to compensate for the reduction in stored energy. In practice, it becomes increasingly difficult to endure optimum conditions of gap setting as power is increased because

(i) The dielectric gets contaminated with metal particles and breakdown will occur at lower voltage.

(ii) With increasing power, R is reduced. This helps in dielectric breakdown at lower voltages.

Relaxation circuit with series inductance The presence of inductance in the relaxation circuit ensures that at the beginning of the charging cycle the current is zero. The rate of change of voltage across the capacitor and the gap is also zero. This voltage, tending to reform the discharge across the gap, is initially slow to rise, and hence, more time is available for deionization of the dielectric, and higher sparking frequencies are possible. Again, if the inductance is such that the circuit is oscillatory, a further increase in charging efficiency and power is possible. It has been reported that the modified circuit can increase the machining efficiency by 25 to 30 per cent and machining speed can be increased by about four times that attainable with the R-C circuit.

The main advantage of the relaxation circuits with or without inductance is that they are simple, cheap and rugged. Fairly high frequency pulses can be generated (up to about 10,000 Hz) at low power outputs. With the relaxation circuits, the metal removal rate under ideal conditions is limited to 2g of steel per minute.

Rotary Pulse Generators

In order to increase the metal removal rates, motor generator sets have been developed to supply the required machining power in EDM. These generators are commonly referred to as the rotary pulse generators and produce assymetric output waves so that the advantage of the equivalent of a D.C. power supply can be maintained. The basic circuit of a rotary pulse generator is given in Fig. 4.22. During operation, the capacitor 'C' is charged through the diode 'D' on half cycle. On the following half cycle, the sum of the voltage from the generator and charged capacitor is applied to the gap.

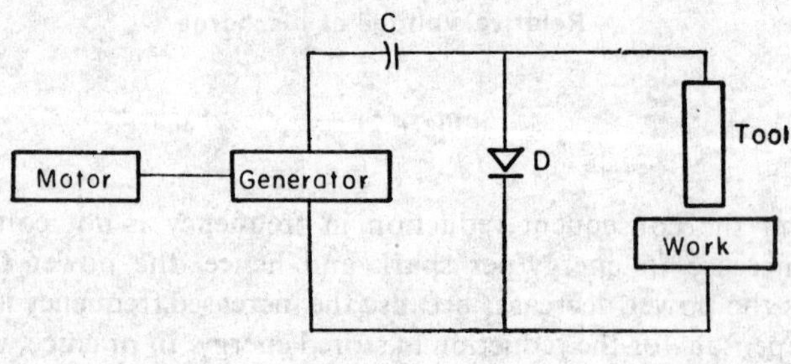

Fig. 4.22

This circuit allows a standard high frequency A.C. generator to be used to produce unidirectional pulses. This circuit permits high metal removal rates but produces exceedingly rough surfaces.

Controlled Pulse Circuit

In the electrical circuits discussed earlier, the switching device was a primary factor in determining the frequency and the amount of energy per discharge. One element of control lacking in the basic circuit is the ability to cut off the current in case of a short circuit. The method of breaking the short circuit in both relaxation and pulse generator circuits is to withdraw the electrode mechanically. However, this takes time and could lead to intensive work surface damage.

The need for a faster method of stopping the current in the event of short circuits resulted in the development of circuits with electronic tubes and transistors. These circuits are known as controlled pulse circuits (static pulse generator) and offer the advantage of faster rate of metal removal and low tool electrode wear.

The majority of spark erosion machines currently available employ transistorized pulse circuits which can achieve higher metal removal rates together with a high degree of accuracy. The use of controlled pulse generators enables wide variations in pulse duration frequency and in the intensity of spark discharges, and employs power transistor triodes as switching devices. These are switched by low power square wave generators and allow independent 'on' and 'off' controls, and the whole system provides unidirectional square-wave pulses to the electrodes.

A vacuum tube circuit is shown in Fig. 4.23. In this case, the resistor R in the R-C circuit is replaced by a series of vacuum tubes connected in parallel. The electronic control circuit BB turns on the tube and the condenser gets charged. This also enables the current flow to stop in case of a short circuit.

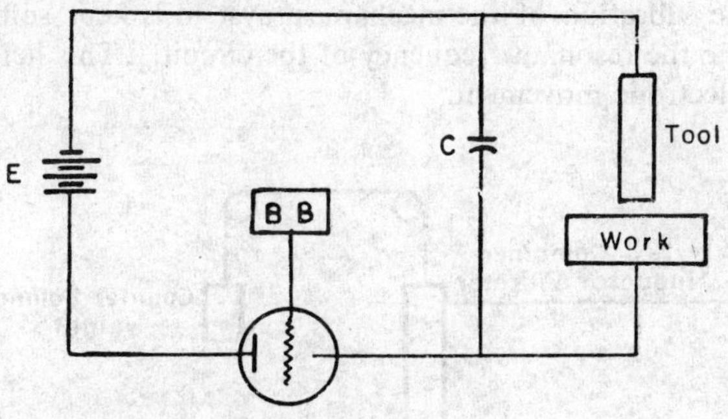

Fig. 4.23

The vacuum tube circuits require high voltages and low currents. Further improvement is possible by the use of transistors in place of vacuum tubes which are low voltage devices. Figure 4.24 illustrates a transistor circuit. In this circuit, switching is done by the oscillator at a selected imposed frequency and does not require the use of capacitors. The oscillator is also controlled by the gap conditions so that the transistors can be turned off in case of a short circuit.

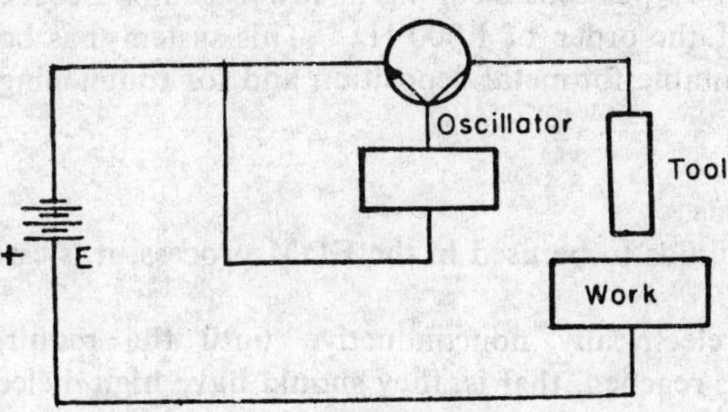

Fig. 4.24

Vibrating Electrode System

If the tool electrode is made to vibrate so that the gap between the tool and work opens and closes regularly, discharge is no longer entirely dependent upon the gap conditions. Sparking will occur when the electrodes are closest, and deionization of the dielectric fluid can take place when the gap is large enough to prevent sparking.

The efficiency of a vibrating electrode could be enhanced when it is synchronized with the resonant frequency of the charging capacitance. Figure 4.25 shows the circuit diagram of the vibrating electrode system with D.C. supply. Due to low circuit resistance, the system operates very near its natural resonance frequency, and the capacitor voltage is virtually twice that of the supply with a corresponding increase in power and charging efficiency. The inductor core and vibrator are combined and the natural frequency of the vibration of the mechanical system is kept sufficiently high as compared to the resonant frequency of the circuit. This helps to obtain the desired electrode movement.

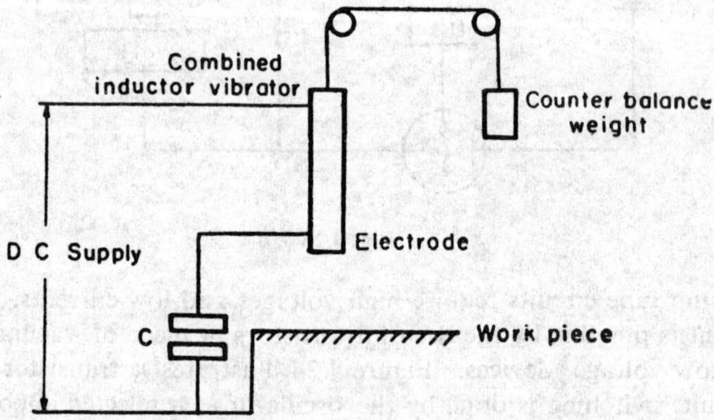

Fig. 4.25   Vibrating Electrode System

The charging efficiency for this type of circuit has been claimed to be of the order of 95 per cent resulting in lower running costs. The vibration frequency is of the order of 1,000 Hz. This system has been found to be particularly suitable for metal deposition and for toughening and roughing operations.

Dielectric Fluids

For dielectric fluids to be used in the EDM process, it is essential that they should

(i) Remain electrically nonconductive until the required breakdown voltage is reached, that is, they should have high dielectric strength.

(ii) Breakdown electrically in the shortest possible time once the breakdown voltage has been reached.

(iii) Quench the spark rapidly or deionize the spark gap after the discharge has occurred.

(iv) Provide an effective cooling medium.

(v) Be capable of carrying away the swarf particles in suspension, away from the working gap.

(vi) Have a good degree of fluidity.

(vii) Be cheap and easily available.

Light hydro-carbon oils seem to adequatey satisfy these requirements. The common fluids that can be used are transformer oil, paraffin oil, kerosene, lubricating oils or various petroleum distillate fractions. Recently, distilled water has also been used in place of dielectric fluid and this has been found to permit very high metal removal rates. Table 4.1 compares the performance of several dielectric fluids on brass tools, when used for machining steels.

Table 4.1

Dielectric fluid	Machining rate Work material removed $cm^3/amp\ min \times 10^4$	Wear ratio*
50 viscosity SSU hydro-carbon oil	39.0	2.8
Distilled water	54.6	2.7
Tap water	57.7	4.1
Tetraethylene glycol	102.9	6.8

$$*\text{Wear ratio} = \frac{\text{volume of work material machined in unit time}}{\text{volume of electrode material worn out in unit time}}$$

The dielectric should be filtered before re-use so that the contamination of the dielectric fluid will not affect machining accuracy. This is usually accomplished by filtration.

Flushing

Flushing is defined as the correct circulation of dielectric fluid between the electrodes and workpiece. Suitable flushing conditions are essential to obtain the highest machining efficiency. In order to comprehend the importance of correct flushing in EDM, it is necessary to understand the phenomenon that occurs in the machining gap when flushing is absent.

To start with, the dielectric is fresh, that is, it is free from eroded particles and carbon residue resulting from dielectric cracking, and its insulation strength is high. With successive discharges the dielectric gets contaminated, reducing its insulation strength, and hence, discharge can take place easily. If the density of the particles becomes too high at certain points within the gap, 'bridges' are formed which lead to abormal . discharges and damage the tool as well as the work electrode. This build-up of the wear debris is eliminated by 'flushing'. Thus, flushing in EDM is as important as any of the electrical parameters and should be adjusted to give that degree of contamination in the dielectric which yields optimum results.

Flushing in EDM can be achieved by any one of the following methods:
  (i) Injection flushing.
  (ii) Suction flushing.
  (iii) Side flushing.
  (iv) Flushing by dielectric pumping.

Injection Flushing

The dielectric fluid is injected continuously into the working gap either through the work piece or tool. A hole is provided in the work piece or tool for this purpose.

Suction Flushing

In this method, the fluid is sucked either through the work piece or the tool electrode. Compared with injection flushing, suction avoids taper effects due to sparking via particles along the sides of the electrode. Suction flushing through the tool rather than through the work piece has proved to be more efficient.

Side Flushing

When flushing holes cannot be drilled either in the work piece or the tool this type of flushing is employed. For the entire working area to be evenly flushed, special precautions have to be taken for the pumping of dielectric.

Flushing by Dielectric Pumping

Flushing is obtained by using the electrode pulsation movement. When the electrode is raised, the gap increases, resulting in clean dielectric being sucked into mix with contaminated fluid, and as the electrode is lowered, the particles are flushed out. This method has been found particularly suitable in deep hole drilling.

Electrodes for Spark Erosion

The spark erosion process is basically a copying process and the shape and accuracy of the machined part will, therefore, depend primarily on the shape and accuracy of the tool or cutting electrode. The spark erosion machine should ensure that under identical conditions, the maximum accuracy in size and shape is obtained. No matter how accurate the machine is, the work produced can never be more accurate than the electrode that machines the work.

It has been observed that in the erosion process, both the work piece as well as the electrode get eroded. Hence, the accuracy of the machined part obtained depends on the electrode wear.

The amount of erosion suffered by the electrode compared with that by the work piece is referred to as wear ratio and depends on the physical and chemical properties of both the electrode and work piece material and the environmental conditions. The melting point of the tool electrode material has been found to affect its wear rate and has been found to follow the law:

$$R_t = 10.15 \times 10^3 \, M_T^{-2 \cdot 28}$$

where R_t = average metal removal rate from electrode (tool) ($\mu m^3/amp \, min \times 10^4$) and,

M_T = melting point of the tool material.

Also it has been shown that

$$W_R = 2.25 \, M_r^{-2 \cdot 3}$$

where W_R = wear ratio (work/tool), and

M_r = melting point ratio (work/tool)

Apart from the melting point ratio, other factors which also influence the wear ratio are:

Metal removal rate Generally the wear ratio increases with cutting rate.

Cross-sectional area of the electrodes Wear ratio is reduced with increase in cross-sectional area.

Work piece material Sintered hard-metals induce considerably higher wear ratio. High alloy steels containing vanadium or molybdenum result in somewhat higher ratio than is commonly encountered with normal steels and nimonic alloys.

Configuration of the electrode A form with slender projections or a narrow section will exhibit a higher tool wear than one of a similar cross-section which is more compact. Thus, a star section will show more wear than a circular section.

Depth Deep cavities or long trough forms will provide a higher electrode tool wear than shallow ones. The reason for this is increased difficulty in swarf removal and proper coverage of the operating zone with dielectric liquid; these factors interfere with the correct spark machining function.

Selection of Electrode Material

Four main factors determine the suitability of a material for use as an electrode. These are:
  (i) The maximum possible metal removal rate.

   (ii) Wear ratio.

  (iii) Ease with which it can be shaped or fabricated to the desired shape.

  (iv) Cost.

From purely technical considerations, it is possible to specify a material, such as silver-tungsten alloy as the most efficient electrode providing a high metal removal rate and very high wear ratio, but the cost of such an electrode under most conditions is, of course prohibitive. Generally speaking, by using a sufficient number of electrodes of material having a low wear ratio, it is possible to produce the same accuracy of machining as with a single electrode of material with a high wear ratio. Keeping in mind technical and economic considerations various material that can be used as tool electrodes are given in Table 4.2. However, it has been found that the major controlling factors of wear ratios, metal removal rates and cutting stability are the functions of the power supply circuit; for this reason, it is impossible to provide a fixed set of rigid rules for electrode section.

Table 4.2

Selection of Electrode Material

Material	Wear ratio	Metal removal rate	Fabrication	Cost	Application
Copper	Low	High on rough range	Easy, can be sprayed also	High	On all metals
Brass	High	High only on finishing ranges	Easy	Low	On all metals
Tungsten	Lowest	Low	Difficult	High	Only where small holes are to be drilled
Tungsten copper alloys	Low	Low	Difficult	High	Used for higher accuracy work
Cast iron	Low	Low	Easy	Low	Can be used only on few materials
Steel	High	Low	Easy	Low	Can be used for finishing work only
Zinc based alloys	High	High on rough ranges	Easily die casted	Low	Can be used on all metals
Copper graphite	Low	High	Very delicate and hence difficult	High	Can be used on all metals

Tool Electrode Design

The tool electrode must be designed as a mirror image of the work to be produced. However, a certain amount of clearance should be provided between the tool and work cavity produced. The magnitude of the clearance varies with the rate of metal removal, the materials of the tool and work. Different tools may be needed for rough and fine machining.

Tables 4.3 and 4.4 show the effect of operating conditions on side clearance during boring.

Table 4.3

Effect of Operating Condition on Side Clearance

Rate of cutting	Finish	Side clearance mm
Rapid	Coarse	0.5-0.6
Medium	Medium	0.2-0.3
Very slow	Fine	0.03-0.06

NOTE: Work-NiCr steel; Tool-brass (50 per cent Cu).

Table 4.4

Effect of Tool Material on Side Clearance
at Very High Rate of Metal Removal

Tool material	Side clearance mm
Brass (60 per cent Cu)	0.25
Duralumin	0.30
Copper graphite	0.35

NOTE: Work-hardened carbon steel.

Surface Finish

The surface produced by the EDM process consists of a multitude of small craters randomly distributed all over the machined face. The CLA value of the surface finish in this case ranges between 2 and 4 μ. The quality of surface mainly depends upon the energy per spark. If the energy content is high, deeper craters will result, leading to a poor surface (Fig. 4.26). The surface roughness (H_{cla}) has also been found to be inversely proportional to the frequency of discharge.

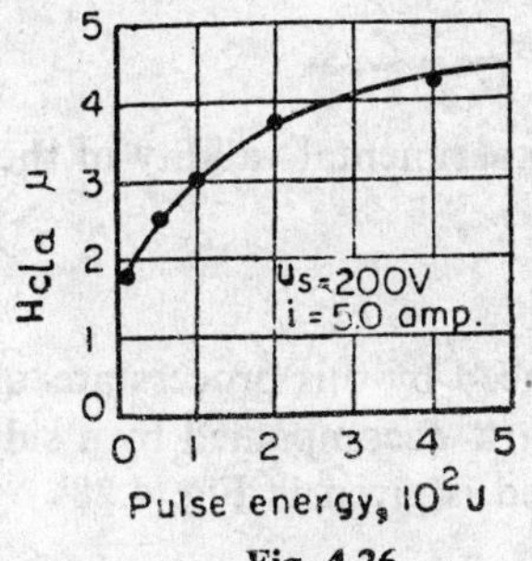

Fig. 4.26

Assuming that each spark leads to a spherical crater formation on the work surface, the volume of metal removed per crater will be proportional to the cube of the crater depth.   Also, it is assumed that

$$H_{cla} \propto h$$

where

H_{cla} = centre line average value of the surface produced, and
h = maximum crater depth.

Also

$$H_{cla} \propto 1/f$$

Therefore

$$H_{cla} \propto h/f$$

Or

$$H_{cla} = K_1 \, h/f$$

where K_1 = constant of propotionality.

The volume of metal removed per discharge (V_1) will be equal to the volume of crater produced.
Therefore

$$V_1 = K_2 \, h^3 = K_3 \, V_o^2 \, C$$

where K_2 and K_3 are constants.

But

$$h \propto H_{cla}$$

Therefore

$$H_{cla} = K_4 \, V_o^{2/3} \, C^{1/3}/f$$

The frequency

$$f \propto R^{-m} \, C^{-n} \text{ (for an R-C type circuit)}$$

Therefore

$$H_{cla} = K_5 \, V_o^{2/3} \, C^{1/3+n} \, R^m$$

Experimental investigations in a majority of cases give a relationship of the type.

$$H_{cla} = K \, V_c^{0.5} \, C^{0.31\sim0.36}$$

Figure 4.27 shows the experimental validity of the above relationship.

Machining Accuracy

Taper   The holes produced by this process are usually tapered due to the presence of a frontal spark accompanied by a side spark.   An exaggerated view of the hole produced is given in Fig. 4.28.

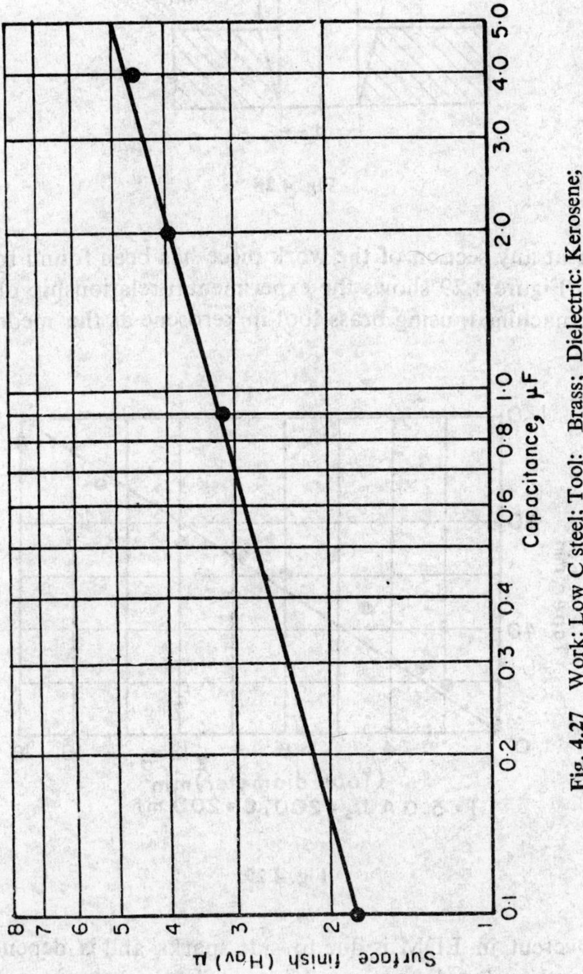

Fig. 4.27 Work: Low C steel; Tool: Brass; Dielectric: Kerosene; Current; 5 amp; Voltage: 200 V.

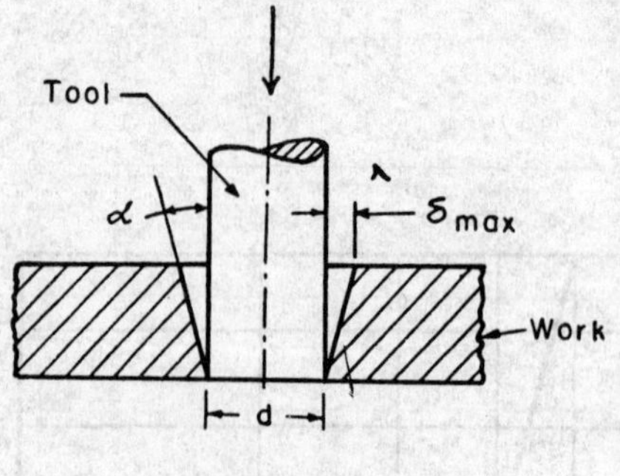

Fig. 4.28

The taper at any section of the work piece has been found to be proportional to d^2. Figure 4.29 shows the experimental relationship obtained when carbon was machined, using brass tool in kerosene as the medium.

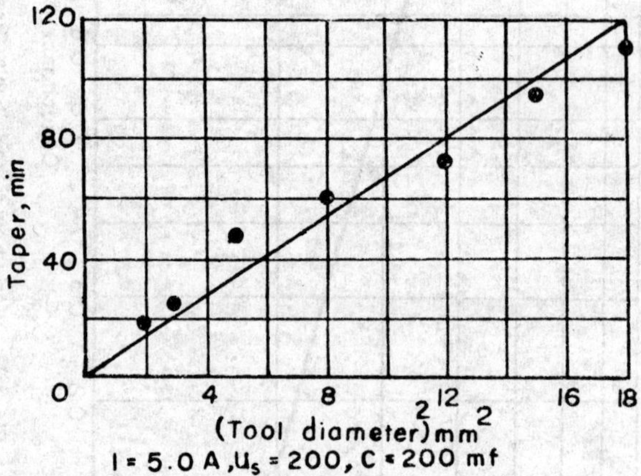

Fig. 4.29

Overcut   Overcut in EDM is due to side sparks and is dependent on the gap length and crater dimensions.   Lazarenko has shown experimentally that overcut O can be expressed by the relationship

$$O = AC^{1/3} + B$$

where A and B are constants, the values of which depend upon the tool work pair.   Table 4.5 gives typical value of A and B.   Dependence of the overcut on the capacitance C is shown in Fig. 4 30.

Table 4.5

Tool work	A $U_b = 100$ V	B $U_b = 100$ V
Copper-copper	0.032	0.015
T 15 K-60-copper	0.032	0.015

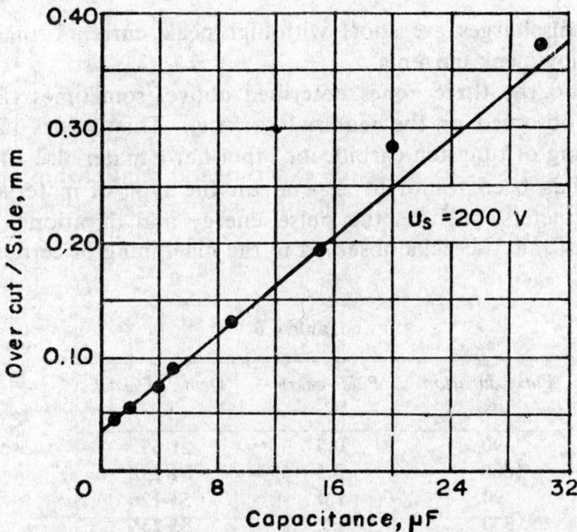

Fig. 4.30 Dependence of overcut on capacitance
Tool: Brass 10 mm diameter; Work:
low C steel

Characteristics of Spark Eroded Surfaces

In EDM, material removal is principally due to thermal phenomenon and local temperatures in the region of 8,000 to 12,000°C are likely to develop. This temperature will have an effect on the structure and the mechanical properties of machined surfaces. The effect may or may not be significant, depending upon the type of work material and the working conditions employed.

A typical cross-section of a steel specimen after machining by the EDM process, when examined, would normally exhibit three different regions (Fig. 4.31).

Region 1 A layer of molten metal, ejected and partly redeposited.

Region 2 Recast metallic layer usually referred to as white layer. The layer has no fixed thickness and is very hard.

Region 3 An annealed layer. Thickness of the annealed layer depends upon the energy of discharge. It has also been found that the zone is

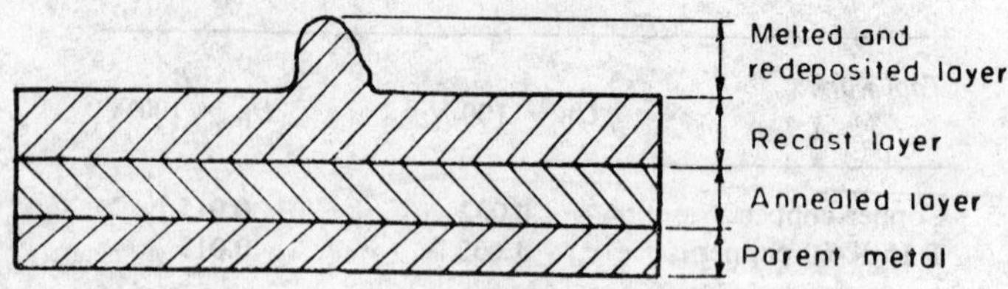

Fig. 4.31

thinner if the discharges are short with high peak currents than if they are long with low peak currents.

In addition to the three zones described above, sometimes tiny micro-cracks can be observed on the material surface. This occurs particularly in the machining of tungsten carbide or other hard materials. The size of micro-cracks has been found to depend on the type of material and the electrical parameters, such as the pulse energy and duration. Table 4.6 gives the size of micro-cracks observed in the machining of cermets.

Table 4.6

Pulse duration μs	Pulse energy 'J'	Depth of crack μ
90	0.5	21-63
1000	0.5	84-252
90	1.0	53-126
330	4.5	85-235

Generally speaking, the crack depth increases with pulse duration and energy.

Machine Tool Selection

A variety of EDM machines ranging from small machines to large units, are now commercially available. The factors that have to be considered in their selection are the (i) number of parts to be machined, (ii) accuracy required, (iii) size of the workpiece, (iv) depth of the cavity, and (v) orientation of the cavity.

Equipment must be versatile and accurate, for tool room work where a variety of workpiece configuration is encountered, EDM machine tool design and construction is a function of the accuracy required. In cases where the positioning accuracy need not be held closer than 0.025 or 0.050 mm, a conventional coordinate table can be used to obtain the position read-out from the lead screw via the hand wheel dial. For higher accuracy, an optical read-out independent of the lead screw is desirable.

Large sized jobs require machines with high rigidity to avoid excessive deflection. High rigidity is also essential whilst working with large sized electrodes. The electrode holding column must be made rigid enough to support the weight of the electrode and also to withstand the coolant back pressure, a peculiarity of this process.

Applications

Spark machining is used for the manufacture of tools having complicated profiles and for a number of other components. The decision to use the spark erosion process for either of these broad applications is usually based on one or more of the basic characteristics inherent in the process.

Spark erosion provides an economic advantage for making stamping tools, wire drawing and extrusion dies, header dies, forging dies, intricate mould cavities, etc. It has been extensively used for machining of exotic materials used in aero-space industries, refractory metals, hard carbides, and hardenable steels.

Delicate work pieces, such as copper parts for fitting into vacuum tubes, can be produced by this method. The workpiece in this case is too fragile to withstand the cutting tool load during conventional machining.

Sometimes accuracy requirements dictate the use of EDM for two main reasons:

 (i) When repetitive shapes are required, they can often be produced from an easy-to-make male electrode.

 (ii) When machining accuracy must be maintained after heat-treatment of the part.

Future Trends

Spark machining has been in production use for approximately seventeen years and is a well-established process for producing holes and cavities in tough materials with high precision.

There are, however, quite a number of problems still to be solved to enable the process to be adopted on an extensive basis. The problems are:

 (i) Ignition of spark discharge in a contaminated flushed dielectric.

 (ii) Mechanism of metal erosion by spark discharge.

(iii) Distribution of discharge energy on anode and cathode during a pulse and its relation to the discharge, conditions width of gap and electrode material.

(iv) Distribution of temperatures and pressures in the discharge gap for different operating times.

There are many instances of spark erosion being applied as a production process in addition to its more usual capacity as a jobbing process. Its use as a production process is largely the result of multiple tooling techniques which allow several components to be eroded simultaneously. A prominent factor of multiple tooling is the development of multi-channel techniques. Work has been conducted with as many as 77 channels and it is highly probable that there will be a considerable increase in the number of channels used.

Numerical control has been tried in the case of a few machines but the technique is still in its infancy.

PLASMA ARC MACHINING (PAM)

Plasma arc was initially employed to cut metals that are difficult to machine by conventional methods. However, in recent years, plasma arc has also been successfully used for spraying, surfacing and welding metals like aluminium, stainless steel, titanium, etc.

Plasma

When heated to elevated temperatures, gases turn into a distinctly different type of matter which is plasma. The changes that take place when gases are heated to a few thousand degree are:

(i) the number of collisions, elastic or inelastic, between atoms increases.

(ii) the gas ionizes, so that a portion of the atoms are stripped off their outer electrons, resulting in the creation of electrons and ions.

(iii) the electrons thus produced, in turn, collide with atoms, heat them through relaxation processes so that their thermal kinetic energy increases, excite them so that de-excitation light is emitted from the atoms and ionize them so that more electrons and ions are produced. Thus, the new matter is characterized by its ability to conduct electricity due to the presence of free charges. At high charge densities, the matter also becomes bright due to emmission from atoms.

The plasma is encountered in electrical discharges, such as flourescent tubes and electric arcs, lightnings, high temperature combustion flames and the sun. In the context of metal working emphasis is on electrical discharges, particularly in electric arcs.

Nonthermal Generation of Plasma

The method of heating the gases by first ionizing them is one of the most popular methods of generating hot plasma. This can be done either by applying a suitable electric field across the gas column or by exposing the gas to ionizing radiation; the latter method being very uncommon.

When gases are heated by an applied electric field, an igniter supplies the initial electrons, which accelerate in the field before colliding and ionizing the atoms. The free electrons, in turn, get accelerated and cause further ionization and heating of the gases. The avalanche continues till a steady state is obtained in which the rate of production of the free charges is balanced by recombination and loss of the free charges to the walls and electrodes.

The actual heating of the gas takes place due to the energy liberated when free ions and electrons recombine into atoms or when atoms recombine into molecules. The bonding energy thus liberated is manifested in the form of kinetic energy of the atom or molecule formed by the recombir.ation. It is thus clear that the heat content of molecular gases is higher than that of mono-atomic gases.

The energy input for such a machanism is determined by the enthalpy of gas at the required temperature, thermal conductivity of the gas and loss of energy in the form of radiation. The electrical conductivity of the gas is a relevant parameter for ascertaining the current density in the plasma for the given field.

The principle of plasma generation is shown in Fig. 4.32. In this case, the high velocity electrons of the arc collide with the gas molecules and produce dissociation of diatomic molecules followed by ionization of the beam. The plasma forming gas is forced through the nozzle duct in such a manner as to stabilize the arc. Much of the heating of the gas takes place in the constricted region of the nozzle duct, resulting in relatively high exit gas velocity and very high core temperatures up to 16,000°C.

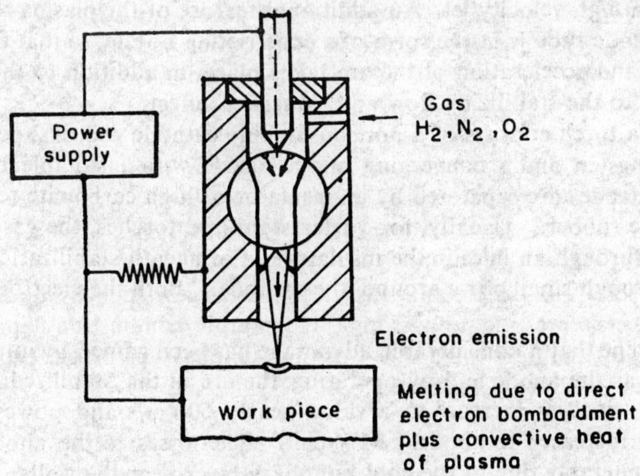

Fig. 4.32

Mechanism of Metal Removal

The metal removal in PAM is basically due to the high temperature produced. The heating of the work piece is, as a result of anode heating, due to direct electron bombardment plus convection heating from the high temperature plasma that accompanies the arc. The heat produced is sufficient to raise the work piece temperature above its melting point and the high velocity gas stream effectively blows the molten metal away. Under optimum conditions, up to approximately 45 per cent of the electrical power delivered to the torch is used to remove metal from the work piece.

Plasma arc cutting was primarily employed to cut metals that form a refractory oxide outer skin. The arc heat is concentrated on a localized area of the work piece and it raises it to its melting point. The quality of cut is affected by the heat flow distribution; uniform heat supply through-out the thickness of the material produces a cut of excellent quality.   The optimum cutting speed (for a given torch) is dependent upon the uniform distribution of heat flow at the plasma to material interface.

PAM Parameters

The parameters that govern the performance of PAM can be divided into three categories:
 (i) Those associated with the design and operation of the torch.
 (ii) Those associated with the physical configuration of the set up.
(iii) Environment in which the work is performed.

The Torch

The modified stabilized arcs came to be known as plasma torches or plasma jets because the plasma of the arc column is pushed out of the nozzle in the form of a high velocity jet. An additional feature of the plasma torches is that the anode tube is in the form of a constricting nozzle, so that further confinement and acceleration of the arc takes place, in addition to the con-striction due to the stabilizing flow and 'Maecker' effect.

The plasma torch consists of a nonconsumable cathode rod of 2 per cent thoriated tungsten and a converging anode nozzle with a suitable orifice. The two electrodes are separated by an insulator of high carbonate resin or some suitable rubber.   Usually, for vortex stabilized torches, the gas is fed tangentially through an inlet in the insulator.   For sheath stabilization, the gas is fed through small ports around the cathode.   Both the electrodes are water-cooled.

It is apparent that a considerable advantage has been gained by introduc-ing a nozzle as the anode and by operating the arc in the stabilized mode. The plasma velocities increase to a value nearly 500 m/s and power den-sities can be increased to about 1 MW/cm^2.   The losses to the anode are reduced considerably due to the cold boundary layer near the wall.

This aspect of high velocity and high power density achieved in the plasma torches is widely used in high temperature work.   The plasma jet has free access, unlike the arc column, between two electrodes, so that the jet can be used as a heat source for all materials, metals and insulators alike.   The plasma flame obtained has temperatures much higher than that achieved in conventional flames (the highest temperature achieved is 3800°C in oxy-actylene flame, as compared with 10,000°C in plasma torches). High power density, high temperature and high velocity jet can be used for spraying and cutting.   The control provided by gas flow and power input enables the use of these torches for a variety of applications in welding, spherodizing chemical synthesis, etc.   The torches also provide the only simulating source for space research experiments.

Modes of operation of D.C. plasma torches   Two modes of electrical con-nection in the torch are used.

Nontransferred arc mode The D.C. power source is connected directly across the cathode and the nozzle, so that the cathode and nozzle carry same current. In such cases, the plasma is in the form of a flame, useful for spraying, ceramic working and chemical synthesis. The hottest portion does not appear outside the nozzle. The anode dissipation is lost in useless heating of the nozzle. The electrothermal efficiency is about 65 per cent for sheath stabilized torches and 75 per cent for vortex stabilized torches.

Transferred arc mode In this mode of operation, the cathode is connected directly to the negative of the D.C. source, while the anode nozzle is connected to the positive of the supply through a suitable resistor to limit the current through the nozzle to about 50 amp. The metal workpiece to be processed is then connected directly to the positive of the supply. When ignited, a pilot plasma flame is established between the cathode and nozzle which provides a conducting path for a high current constricted arc between the cathode and workpiece. Once this arc is struck the pilot flame circuit is disconnected. This method is limited to cutting, welding and hard surfacing of metals.

The transferred arc mode utilizes the anode dissipation for useful heating of the workpiece, and in this way, the effective electrothermal efficiency is increased to 85-90 per cent, the only loss being at the cathode and the unconnected nozzle.

Both these modes of operation use argon or nitrogen or a mixture of the two. For certain useful purposes, a percentage of hydrogen may be added. The torches are operated on two flow modes.

Turbulent mode When high velocity flames are required for material removal by melt blasting or spraying, high gas flow rates, which give rise to turbulent jets, are used. These flames are short in length and are rather cold outside the nozzle. These are used for cutting, low quality welding and spraying.

Laminar mode When low velocity or long flames are required, this mode of torch operation is used. Low gas flow rates are maintained in long nozzles to obtain a laminar flame. This is used wherever the processed material is not to be sputtered out or when the disintegration of molten drops into fine droplets is undesirable. Flames with velocities 50 m/s and lengths up to 900 mm can be achieved.

Design of D.C. plasma torches The plasma torch is designed to obtain maximum thermal output. The increase in efficiency not only helps in achieving better heating of the gas but also in reducing electrode losses and thereby increasing the life the of electrode. The design also ensures that the erosion rates of the electrode are kept to a minimum.

The parameters which affect the performance of the torch are cathode size, its taper near the gap, convergence of the nozzle, nozzle orifice diameter and orific length, electrode gap and cooling of the electrodes.

The cathode The cathode taper at the tip and diameter of the cathode affects the drop across the cathode rod; the diameter and the bluntness or sharpness of the cathode determines the erosion rate, stability of the arc and flame shape. Tapered rods with slightly blunt tips are used for non-transferred applications, while for cutting applications, a flat disc sloped on the edges is used. The taper angle for N_2 is kept larger than that for argon (about 45°). The taper angle or sloping of the disc is usually smaller or equal to the convergence angle of the anode nozzle. For spraying and cutting, 10 mm dia. cathodes are used up to 500 amp, while for welding, electrodes of 6 mm dia. are used. The cathode is pressed to fit into a water-cooled copper holder and brazed to it.

The anode nozzle The anode convergence has a definite relation with the operation of the torch and this is matched to the cathode taper to obtain the best performance, that is, minimum electrode erosion and electrode loss.

The nozzle orifice is designed on the basis of the development of a stabilized plasma column. In a particular model, the plasma is divided into two zones, viz. undeveloped flow region and fully developed flow region. In the fully developed zone, the electrode losses are higher. Hence in order to obtain minimum anode loss and erosion, it is preferable that the plasma flame be out of the nozzle duct before it becomes fully developed. Theoretically, the nozzle length, diameter and electrode gap can be derived. The usefulness of such criterion is limited only to cutting torches where the application does not impose any additional restriction. But spraying, welding and laminar flow torches override this criterion because they require long throat lengths. Hence, in these torches, the ruling criterion of a long nozzle gives rise to increased anode loss. In transferred arc torches, the throat length is also limited with a view to prevent double arcing and short arcing.

The combination of the electrode gap and anode orifice is optimized for a given enthalpy of the flame, and therefore, for a given power input and arc current.

The nozzle diameter determines the power density in the flame. Obviously for cutting applications which require highly constricted flames of high velocity, nozzle diameters are kept to a minimum value compatible with minimum electrode erosion and minimum double arcing. However, in spraying and other applications, nozzle diameters can be varried to obtain various velocities and power densities. The minimum and maximum diameters of the nozzle are determined by the arc current, as is the cathode diameter.

It is also seen that these parameters are governed by the gas flow rate electrode cooling, open circuit voltage and load voltage.

With so many parameters, it is necessary to experimentally obtain the various dimensions for a particular operating condition. The various dimensions for different types of torches are given in Table 4.7.

General considerations The general guidelines for designing the torch are:

(i) Large currents require a larger orifice diameter cathode, orifice length and electrode gap.

(ii) While operating at larger arc voltages, it is preferable to increase the angle of taper at the cathode tip. By the same token, while operating in the transferred arc anode (high arc voltage), flat face cathodes are used.

(iii) For avoiding turbulence, the edges of constructions are rounded off and smaller cathode diameters are used. To stabilize laminar flow modes, a large orifice length/diameter ratio is used.

(iv) In any torch, the alignment between the electrodes is very essential and any misalignment would lead to undesirable arcing on the sides and non uniform erosion on one side of the electrodes.

(v) Non transferred arc modes use long throat lengths, while for transferred arc mode, the minimum orifice length is used.

(vi) The cathode taper and convergence of the nozzle should be such that the arc terminates at the tip of the cathode. This usually means that the tip is the nearest point to the anode. In some cases, the tapered sides of the cathode are made parallel to the converging side of the nozzle.

(vii) The insulator portion should not be placed anywhere near the arc zone to avoid any damage to the insulator by radiation or convection heat.

(viii) The cooling provided to the cathode should be optimized for minimum uniform erosion without cracking the cathode and destroying the maximum stability. But excessive cooling in rod type electrodes at normal cooling rates is not usually obtained.

(ix) High quality insulation should be used in other portions of the torch as high frequency voltage is used for ignition.

(x) Oxygen or compounds of oxygen in the plasma gas are detrimental to the cathode and by the same token, no water leakage into the plasma chamber from the chamber cooling the electrodes should be allowed.

Types of Torches
(i) *Non-transferred Arc Torches*
(a) *Turbulent mode flame torches* In these torches, the flame obtained is short in length (about 15 cm at 400 Amp in nitrogen), of high velocity and is emitted with a characteristic 'shriek' or 'hiss'. Such torches use rod type cathodes and nearly 25 mm nozzle throat lengths and the orifice diameters can be varied over a wide range. These torches are used for spraying, insulator working and chemical synthesis.

In certain torches, an inert gas flow surrounding the main flame is

TABLE 4.7

Various Dimensions and Operating Conditions for Different Torches

Type of torch	Cathode dia.	Cathode tip	Nozzle dia.	Nozzle length	Electrode gap	Power input	Gas flow rate
Spray torch	8-12 mm rod type	Taper 30-60° with a tip 1 mm	4 to 10 mm	10 to 25 mm	1 to 5 mm	5 to 40 kW	1 to 30/min
Laminar sphero-dising torch	6-10 mm rod type	Taper 30° with a tip 1 mm	6 to 10 mm	25 to 100 mm corner should be rounded	3 to 5 mm	5-30 kW	1 to 101/min
Cutting torch	10-20 mm disc type	Taper 75° with a tip 8 mm	3 to 5 mm	3-5 mm	1 to 4 mm	10-200 kW	20 to 60 1/min
Welding torch	3-6 mm rod or disc type	Taper 75° with a blunt tip	3 to 4 mm	3 mm (corner should be rounded)	1 to 3 mm	5 to 50 kW	5 to 20 1/min
Laminar welding torch	6-8 mm rod type	Taper 30° with a tip 1 mm	3 to 4 mm	25 to 100 mm (corner should be rounded)	1 to 3 mm	5 to 50 kW	2 to 10 1/min
Micro torch	2 mm rod type	Taper 45° sharp tip	0.5 to 2 mm	2 to 5 mm	1 to 3 mm	upto 2 kW	5 to 2 1/min

provided to shield the substrata from the atmosphere.

(b) *Laminar mode flame torches*   Such flames have low velocities, lengths as long as 1 metre and the flame is emitted with a soothing hum. The cathode diameters are amall and the nozzle throats (up to 125 mm) are large. The flame is used for spherodizing and melting ceramics.

(c) *High power torches*   These torches are made for high temperature arc tunnels and are operated at very high currents (2,000 amp or higher). To avoid electrode erosion, special magnetic confining fields are used at the nozzle.

(ii) *Transferred Arc Torches*
(a) *Cutting torches*
(1) *Single flow torch*   The torch is of the usual type, with the cathode in the shape of a disc, tapering on the sides. The nozzle throat length is kept minimum (3 to 5 mm). High gas flow rates are maintained to obtain high velocities. The transferred arc is used for cutting steels of all types, aluminium and sometimes copper.

(2) *Dual flow torch*   An additional gas flow surrounding the main arc is provided. This gas flow shields the workpiece. In carbon steel cutting, the dual gas flow steam is of oxygen, so that the plasma cutting is agumented by oxygen cutting to achieve very high cutting speeds. The dual flow has also the advantage of avoiding double arcing and further construction of the arc outside the nozzle due to the surrounding cold flow.

(3) *Multiport nozzle torch*   The dual flow concept is used in a different form by providing part of the plasma gas as shielding flow through small ports surrounding the main orifice of the nozzle. This flow constricts the arc and also shields the workpiece. However, in an ordinary torch, oxygen cannot be use.

(4) *Oxygen plasma cutting torch*   With zirconium as cathode, torches can be used with oxygen plasma for a short duration electrode life. This is used in single flow or in multiport nozzle torches.

(b) *Welding corches*   These torches are operated for minimum turbulence and low velocity so that the molten metal is not thrown away. The cathodes are usually of a smaller diameter and nozzles are longer than that used in a cutting torch. For better performance, laminar transferred arc torches are used (particularly for welding powder compacts, e.g., zirconium).

(c) *Micro torches or needle torches*   These are similar to welding torches, except that they are used at very low powers (1 kW) for welding and cutting thin foils and wires. The torches are used in transferred or nontransferred arcs. It is found that even for obtaining pilot flames at the low current required, high open circuit voltages are necessary.

Cutting Gases
The types and magnitudes of the flow rates of cutting gas depends on

the application. Generally speaking, the thicker the work piece, the greater is the gas flow requirement. The choice of the gas will depend on economics and the quality of the cut edge desired. The different gases produce cut faces of varying degrees of smoothness.

Aluminium and magnesium can be cut with nitrogen and nitrogen-hydrogen mixture or argon-hydrogen mixture. Nitrogen-hydrogen mixtures are used for cutting stainless steel up to 50 mm thick. For heavier sections, a mixture of 65 per cent argon and 35 per cent hydrogen is used.

Carbon steel can be cut most economically with gas containing oxygen. Attempts to use pure oxygen have been unsuccessful because the highly corrosive action of oxygen at plasma temperature rapidly consumes the electrode. The savings realized from using compressed air as the cutting gas are attractive, but the operating life of even the special zirconium insert electrode designed for air cutting is too short for most users.

The most successful approach to the use of an oxidizing cutting gas is to use nitrogen as plasma gas and subsequently introduce oxygen into the plasma downstream of the electrode (dual-flow). This method results in long electrode life and combines the advantages of using inexpensive gas with higher speeds obtained from an oxidizing plasma effluent.

Physical Configuration

In PAM, variables such as torch stand-off, angle, depth of cut, feed and speed of the work towards the torch are important. The feed and depth of cut determine the volume of metal removal. Figure 4.33 shows metal removal as a function of the torch angle.

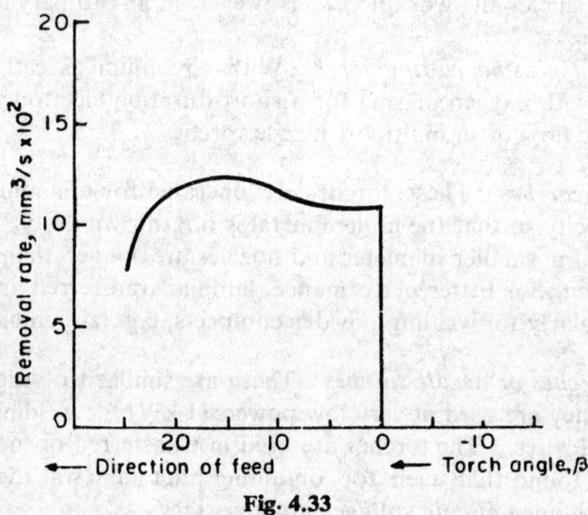

Fig. 4.33

Operation Data

Operation data for aluminium is given in Table 4.9. Conditions for cutting

Table 4.9

Typical Conditions for Plasma Arc Cutting of Aluminium

Thickness mm	Speed m/min	Orifice dia mm	Power kW	Gas flow litre/hour
6	7.5	3	60	2270 Ar—1135 H_2 or 3700 N_2—900 H_2
12	5.0	3	50	1850 Ar—925 H_2 or 3959 N_2—1800 H_2
25	2.25	4	80	1850 Ar—935 H_2 or 3950 N_2—1800 H_2
50	0.5	4	80	1850 Ar—935 H_2 or 3950 N_2—1800 H_2
75	0.375	4.5	90	3700 Ar—2000 H_2 or 3950 N_2—1800 H_2
100	0.3	4.5	90	3700 Ar—2000 H_2 or 3950 N_2—1800 H_2

magnesium are about the same as those for aluminium. Other metals, such as copper, copper-nickel alloys and cobalt base alloys can be cut, using conditions listed for stainless steel in Table 4.10. There are many advantages in the quality of the cut made on stainless steel with a plasma torch as opposed to other types of cutting, particularly oxyfuel gas cutting with iron powder. In cutting with the plasma arc, there is little or no carbide precipitation, very little oxidation, and therefore, little contamination of the metal on the cut face. Magnetic properties of type 304 stainless steel were measured on uncut base metal and plasma cut samples. These tests indicated that magnetic permeability is unaffected by constricted arc cutting.

Stainless steel, mild steel as well as aluminium in thickness up to 25 mm, can be cut at speeds up to 2.5 m/min with best quality occurring at about 1 m/min.

Table 4.11 shows the operating conditions for the plasma arc cutting of carbon steel, using a multiport nozzle.

Table 4.10

Typical Conditions for Plasma Arc Cutting of Stainless Steel

Thickness mm	Speed m/min	Orifice dia mm	Power kW	Gas flow litre/hour
6	5.0	3	60	4250 N_2
12	2.5	3	60	4250 N_2
25	1.25	4	80	5000 N_2
50	0.5	4	100	3700 Ar—2000 H_2
75	0.375	4.5	100	3700 Ar—2000 H_2
100	0.175	4.5	100	3700 Ar—2000 H_2

Table 4.11

Typical Conditions for Plasma Arc Cutting of Carbon Steel

Thickness mm	Speed m/min	Orifice dia mm	Power kW	Nitrogen flow litre/hour	Oxygen flow litre/hour
6	5.0	3	55	4250	2300
12	2.5	3	55	4250	2300
25	1.25	4	85	4250	2300
50	0.625	4.5	110	5700	2850

For mild steel cutting, using dual flow torch with N_2 as plasma gas and oxygen or air as shield gas, the cutting speed for a given thickness is double that of the conventional oxygen torch.

Accuracy and Surface Finish

The amount of thermal energy transferred in the top part of the cut, and hence, also the energy transferred down in the cut is controlled by the cutting speed. If the speed is too high, then the upper edge of of the contact face reaches too far into the plasma jet and receives an excessive amount of heat. The kerf is thus very wide on the top and narrows downwards, forming a V-shaped cross-section. Very low cutting speeds also result in an excessive kerf width at the top; however, due to the excess energycmaining rin the plasma, the kerf widens in the direction of the bottom edge.

The optimum cutting speed is achieved by advancing the torch head at the rate at which the distribution of the heat flow from the plasma into the material is almost uniform throughout the thickness of the material. Under such conditions, perpendicular edges are obtained; that is, the kerf is almost as wide at the top as at the bottom.

The kerf in which the burns are may be considered an extension of stabilizing nozzles. The arc continues to be stabilized by the rapidly flowing relatively cold gases of the boundary zones. There is no well defined anode spot. The contact between the arc and the material is diffused and there is only a certain active anode region in which most of the current passes into the anode. The location of this region depends upon the cutting speed, the intensity at which the arc is blown and the static characteristic of the current source. If the source has a soft characteristic, that is, a sufficient voltage reserve, then the active anode region is easily blown into the lower third kerf. Joule's heat generated by the current flowing through the arc column compensates for all the losses due to the cooling of the arc by the material, while the heat supplied by the anode compensates for the lower amount of heat energy supplied into the bottom third of the kerf.

When thin materials (about 20 to 25 mm) are cut, the high cutting speed shortens the section where the contact surface is perpendicular or slightly inclined. Provided optimum cutting speeds are employed, the slope of the contact surface is smaller than on thick metal and depends on the kind of material to be cut. On stainless steel, the slope angle is 15-20° and on aluminium, 25-30°.

Excessive speeds produce kerf with a marked level because the arc cannot be drawn out to the lower edge of the cut even in thinner material. The cross-section is produced by the current source with a voltage insufficient to lengthen the arc to the requisite value. If the cutting speed is correct and the current source has sufficient voltage, nearly perpendicular and parallel kerf surfaces can be obtained in thin materials. A fused lower edge proves that the active anode region has been kept stable in the lower position, that is, the current carrying plasma is distributed throughout the depth of the cut and thus maintains a uniform heat supply to the material.

In sufficient cutting speed results in excessive heat supply to the region with a lower edge. The material is found over a greater width than the directed plasma jet, and therefore, it is not completely blown out from the cut. As the torch passes on, the molten metal again joins to form bridges.

If two welding sets connected in series are used to cut thicker metal, a moderate widening of the kerf in the middle of the cut face can be avoided.

Plasma arc cuts in plates up to 100 mm in thickness have surface smoothness ranging between that produced by oxyfuel gas cutting and band saw cutting. The cut faces in thicker plates tend to be rougher because of excessive melting caused by lower cutting speeds.

Kerf width of plasma cut is about a half to two times as wide as oxyfuel gas cuts in plates up to 50 mm in thickness. For example, a typical kerf width on a plasma cut in 25 mm stainless steel is about 5 mm. Kerf width increases with an increase in plate thickness. A plasma cut in 175 mm stainless steel made at 100 mm/min produces a kerf width of 30 mm.

'Dross' is the name given to the material that melts during cutting but adheres to the bottom edge of the cut face. Plasma arc cuts are generally dross free if proper cutting conditions and the appropriate cutting gas are used.

Top edge rounding will result if excessive power is used for cutting a given plate thickness. It may occur in high speed cutting of relatively thin material.

Metallurgical Effects
The depth to which the heat effect will penetrate during cutting depends o

cutting speed, or conversely on the duration of time that a given portion of the cut face is exposed to the heat of the arc.   The heat affected zone on the plasma arc cut face of a 25 mm thick stainless steel is only 0.075 mm to 0.12 mm deep.

Work Environment

The environment group of variables for PAM includes any cooling that is done on the cathode, any protective type of atmosphere used to reduce oxidation of the exposed high temperature machined surface, and any means that might be utilized to spread out or deflect the arc and plasma impingement area.   In general, very little has been done to investigate the effects of these techniques for improving the operation, although it is commonly accepted that significant advancements can be made in the processes by optimizing them.

Equipment for D.C. Plasma Torch Unit

For the stable operation of the D.C. plasma torch, the D.C. power source should have a falling characteristic.   The Kaufman criterion is that the slope of the *V-I* characteristic should exceed that of the arc characteristic.

In stabilized arcs, particularly in argon, the arc characteristic has a falling as well as rising characteristic in normal operating ranges.   This, together with the fact that electrode erosion and gas flow rates affect the arc voltage demands that for the stable operation of the arc, the arc voltage fluctuation should not affect the current.   Arc current reduction causes the arc voltage to rise, resulting in further reduction of current regulation.   Thus, it is necessary to have a high degree of current regulation.   This aspect is important in any type of nonconsumable metal-inert gas arc.   The best situation is when the power supply characteristic is vertical.

It is seen that forced operation with a power supply which is not current regulated or which does not have sufficient voltage, leads to improper loading of the torch.   The electrode erosions are increased and the flame extinguishes as soon as the igniting h.f. voltage is topped.   It also results in very high currents being drawn.

While initiating, the cathode and anode are shorted, resulting in a current surge similar to the starting of electric motors.   If a very heavy surge is allowed, then the resulting arc is blown out of the electrode region due to 'Maecker' effect, and consequently, the arc is extinguished.   Thus, the power supply should have a saturation current, beyond which the current should not rise even for a short circuit.

For easy ignition and maintenance of the arc, the open circuit voltage should be higher than the load voltage.   Also, the ripple should be low.

The above factors are admirably satisfied by D.C. motor-generator sets or by transductor controlled rectifier units with saturable core reactors which are more widely used.

The transductor controlled unit comprises a transductor which is connected in series with the main transformer on the input side. Some voltage drops across the transductor inductance which is varied by passing a biasing D.C. current through another winding (the D.C. current saturates the core, reducing the inductance) and the balance appears at the transformer. Thus, the input voltage to the transformer is varied. The current regulation is provided by the fact that the drop across the transductor is proportional to the current drawn, and thus, with increasing current, the voltage at the transformer, and hence, the output voltage drops down. Additional current regulation is provided by passing the load current through yet another winding on the transductor. This limits the short circuit current also.

Selenium rectifiers are preferred to silicon rectifiers because of the surge resistance capacity of the selenium rectifiers. Silicon rectifiers are space saving and wherever they are used, they should be protected by R-C networks for bypassing the voltage surges and by fast acting fuses for current surges. The D.C. power supply specifications for verious types of torches are given in Table 4.12.

Table 4.12

D.C. Power Supply for Various Torches Operating in Argon

Type of torch	Open circuit voltage	Load voltage	Load current
Spray torch	75 V	30 V	600 amp
Laminar spherodising torch	100 V	45 V	400 amp
Cutting torch	500 V	200 V	500 amp
Welding torch	500 V	200 V	300 amp
Laminar torch	600 V	240 V	300 amp

Safety Precautions

It is of utmost importance that the units are operated after ensuring that all health safety measures are observed. The plasma flame emits a highly intense beam, particularly strong in ultraviolet and infrared radiations. These radiations, if taken in large doses, might cause permanent damage to the eye in the form of cataract. Over exposure results in reddening and a gritty feeling in the eyes due to loss of sleep. Over exposure to ultraviolet rays may also cause painful skin burns, and in extreme cases, these lead to cancer. Though these radiations are very intense in all the plasma flames, the laminar plasma flame is much more dangerous. Hence, it is imperative that proper glasses and proper dresses are worn before going close to the torch. The glasses should be good at ultraviolet and infra-red cut off and most of the body should be covered.

It is necessary to operate the torch in an airy room with proper exhaust, as many toxic gases (NO_2, O_2 etc.) are synthesized in the atmosphere. These are very much above permissible levels when proper exhausting facilities do not exist.

The noise level while operating the torch is also very high. Hence, earmuffs or plugs should be used.

Asbestos gloves with an inner layer of leather should be worn for operating hand torches. Health physicians should be consulted regarding the number of hours the operator can remain close to the plasma torch.

In the spraying and chemical synthesis of toxic powders, it is preferable to mechanize the torch and operate it from a distance.

Economics

The following gives a comparison of PAM with oxy-acetylene cutting.

(i) *Plasma Cutting*

(a) *Consumables*

Nitrogen: 4 m³/h	@Re. 1/m³	Rs. 4.00/h
Hydrogen: 0.4 m³/h	@Rs. 2/m³	Re. 0.80/h
Electric power 100 kW	@Re. 0.1/kWh	Rs. 10.00/h
Electrode Replacement		Rs. 5.00/h
Total cost of consumables		Rs. 19.80/h

(b) *Fixed Charges*

Purchase cost of equipment including power sources	Rs. 110,000
Assumed write off duration 5 years or 10,000 h	
Equipment cost	Rs. 11.00/h
Labour and overheads	Rs. 15.00/h
Total of fixed charges per hour	Rs. 26.00/h

From the above, the maximum and minimum hourly operating cost may be assumed as follows:

Min.: Rs. 26/h for zero duty cycle, assuming the operator is nevertheless assigned to the operation.

Max.: Rs. 45.80/h for 100 per cent duty cycle.

(ii) *Oxy-natural Gas Cutting*

(a) *Consumables*

Oxygen 4 m³/h	@Rs. 1.35/m³	Rs. 5.40/h
Acetylene 0.54 m³/h	@Rs. 8.35/m³	Rs. 4.50/h
Total cost of consumables per hour		Rs. 9.90/h

(b) *Fixed Charges*

Purchase cost of torch and tip	Rs. 1500.00
Assumed write off time of 10 years or 20,000 h	
Equipment cost	Re. 0.75/h
Labour and overheads	Rs. 15.00/h
Total cost of consumables	Rs. 15.75/h

Max. : operating cost Rs. 24.975/h.
Min. : operating cost Rs. 15.075/h.
Based on experience, the following conclusion can be drawn:

(1) A single plasma torch, operating at 1 m/min which is a condition for providing very high quality cutting, is lower in operating cost than a two-torch oxygen cutting installation running at an average speed (0.35 m/min) for equivalent duty cycle.

(2) A single plasma torch, cutting at 1 m/min operates approximetely at the same cost as eight torches cutting at an average speed (0.35 m/min) for equivalent duty cycle.

(3) As a rough approximation, under average conditions, plasma cutting can be considered to cost approximately Re. 1/m for material up to 25 mm thick and cutting speeds of 1 m/min.

(4) Cutting cost decreases with increasing cutting speed, hence, plasma arc cutting must be used for high speed cutting operations.

Other Applications of Plasma Jets

Plasma arc process for cutting aluminium and other nonferrous materials was first introduced in 1955. Due to the remarkable results, the process has now been widely accepted by industries for varied applications. The major areas of industrial production where plasma jets have successfully been employed are:

(i) Welding of materials like titanium, stainless steel, etc. which are otherwise difficult to weld.

(ii) Plasma arc surfacing.

(iii) Plasma arc spraying.

Sufficient literature on plasma arc welding is available but little information regarding plasma arc surfacing and spraying exist. This section describes in brief the application of plasma to surfacing and spraying.

Plasma Arc Surfacing

Surfacing is defined as the deposition of filler metal on metal surface to obtain desired properties or dimensions. It is usually employed to extend the life of a part which may otherwise have all the properties necessary for an engineering application, or to replace metal which has worn out or corroded away. The overlay may contribute to corrosion resistance, wear resistance, toughness or anti-friction properties.

When the overlay contributes to abrasion resistance, it is generally referred to as hard surfacing. This term is also applied to overlays used for impact resistance or low friction qualities. Here 'hard' connotes durability.

The plasma arc welding process can be used for fusing thin overlays to base metals at low dilation levels. However, for weld surfacing, metal powder is used instead of wire. The powder is introduced into the effluent of the plasma arc where it is melted and bonded to the surface of the base metal. The deposited metal solidifies as a cast structure, having similar metallurgical characteristics as those of the deposits produced by the gas shielding arc welding process. The plasma arc surfacing process should not be confused with metallizing.

Deposits can be applied from 20 to 50 mm wide in a single pass or can be widened to 80 mm by oscillating the torch. Fused deposits can be applied as thin as 0.25 mm or as heavy as 6 mm in a single pass. Dilution rates can be controlled from 5 to 50 per cent at the deposition rate of less than 0.5 kg per hour to more than 7.5 kg per hour.

The heat distribution afforded by the constricted arc, coupled with the buttering action of the powder, results in overlays having dilution values almost as low as those obtained by the oxy-acetelyne welding process and deposition rates exceeding the gas tungsten arc welding process. A comparison of veld results obtainable by the plasma arc with the weld surfacing process is given in Table 4.13.

Table 4.13

Surfacing Process Comparison Chart

Process	Deposition rate normal range kg/h	Weld dilution normal range per cent	Minimum practical deposition thickness mm	Surfacing material form
Oxy-acetylene	0.5-3.0	1-10	0.8	Rod, powder
Gas tungsten arc	0.5-4.0	10-15	0.8	Rod, wire
Plasma arc weld Surfacing	0.5-7.5	5-30	0.25	Powder
Spray type transfer	4.0-7.5	5-70	1.5	Wire

For economical as well as practical reasons, many hard facing alloys are available only as cast rod or powder. Therefore, their application is generally limited in oxyfuel, gas, tungsten or plasma arc processes. The use of hard facing alloys in powder form allows equipment to be operated continuously and automatically.

Equipment Equipment required for weld surfacing differ from those for welding since the objectives are diverse. Weld surfacing equipment for mechanized operations usually includes a palsma arc torch—a powder dispensing system for storing and accurately metering the surfacing powder to the torch.

Surfacing alloys Many of the hard facing alloys are available in powder form for plasma arc surfacing. These materials can be classed as tin alloy typs with varying degrees of abrasion, impact and corrosion resistances. For best results, surfacing alloy should (i) be sized approximately 150 to 270 mesh, (ii) be metallurgically compatible with the base metal and (iii) have a melting point equal to or less than the base metal to minimize weld dilution.

Plasma Arc Spraying

Spray coating is a process in which a surface of arbitrary thickness is obtained by spraying the previously prepared surface of the base material with droplets of a molten material.

Plasma spraying is a recent development in the field of metallizing and has been gaining wider use as a method for producing surface coating of refractory materials and for fabrication of free standing shapes. The utilization of plasma torches for spraying protective layers has brought about a rapid progress in coating techniques. A higher operating temperature and the possible use of inert atmosphere are the main advantages derived from this innovation.

Spray coating with plasma The technique is similar to that of coating with oxy-fuel or electric arc metallizers. The process consists of two basic operations, the preparation of the surface to be coated and spraying proper.

The surface preparation is of paramount importance for the perfect adhesion of the coating to the parent material. The surface of the parent metal is cleaned by degreasing and appropriately roughened by grit blasting. Degreasing in a hot alkaline bath of trichloetrylene is indispensable if a part has been contaminated with oil. In spraying with plasma guns, grit blasting will usually be sufficient because plasma produces coating with very good adhesion.

The plasma spraying equipment consists of the torch, cooling circuit with a circulating pump, gas cylinders with the pressure regulators and gas supply tubing, storage bin with a device for metering the powder and continuously supplying it into the plasma, D.C. supply source, electrical circuits and gas flow meters. The part of the spraying plasma torch attached to the nozzle also acts as the anode with its front part adapted to mix the powder into the plasma. The aperture for the powder supply must be in a place through which no current flows. The length of the arc depends on the stabilizing gas, the gas flow and the nozzle diameter. In universal torches (for various kinds of stabilizing gases) the cathode is adjustable; thus the optimum length of the discharge path for various gases or gas mixtures can be set without exchanging the nozzle. The transition of the current from the plasma to the individual section of the nozzle-anode can occur from the centre of the plasma as follows: (i) the current can flow through the nozzle wall, (ii) only through the relatively cold layer and (iii) the flow of the current can be mediated by rapid transferred small secondary arcs or diffusion of electrons across the boundary layers.

An important issue in the service life of the nozzle depends on the uniform stressing of its active zones and observance of the permissible stress limits in all sections. Since hydrogen (because of its good thermal conductivity) increases the temperature of the anode, it must not be added in proportion exceeding 10 per cent.

The difference between the plasma spray and oxy-gas spraying can be summarized as follows:

(i) The beam temperature differs. The mean temperature of the plasma beam is about four times that of the combustion gases from an oxy-acetylene gun. Consequently, high melting materials can be sprayed, and generally speaking, even on a less roughened surface of the base material, coating adheres better than it does with oxy-acetylene spraying.

(ii) The discharge rate of plasma is higher than the flow velocity of the combustion gases. This and the higher gas temperature results in a far rapid heat transfer to the particles. Since the latter impinges on the base material, their deformation is more thorough because they are completely molten and have received a higher acceleration. Consequently, the coating is finer and more compact. The higher temperature and velocity of the plasma causes a more efficient utilization of its enthalpy as compared with that of the combustion gases; hence, the plasma torch has a higher spraying output per unit time than the oxy-acetylene gun with the same heat output.

(iii) Selecting the proper atmosphere for the spraying process; materials which are easily oxidized, decarbonized or react with combustion gases of the oxy-acetylene flame can be successfully applied.

(iv) Precision feeding of powder material or wires is more difficult in the plasma torch than in the oxy-acetylene gun. Owing to the higher prime cost of plasma equipment, its cost per unit output is slightly higher, although the direct operating costs of the plasma torch with a nitrogen-stabilized arc are lower than those of an oxy-acetylene gun with the same thermal output.

Control of process variables   Spraying is more difficult with a plasma torch than with a conventional gun; the greater the intricacy of the spraying device, the greater is the number of breakdowns, and the quality of coating is affected by the number of variables. The following factors affect the quality of coating:

(i) Torch-to-work distance affects power dwell time, velocity, impact, density, substrate temperature, deposition efficiency, oxidation, particle adhesion, coat bond strength and stress in the substrate and coating.

(ii) The rate of torch traverse affects the degree of oxidation of the coating sublayers and determines the amount of thermal stress in the coating.

(iii) A 90° torch-to-work angle may not always produce the best results. A slightly off-set position of the torch prevents particle boundback, provides a more desirable directional exhaust and takes advantage of directional surface roughening for better mechanical bonding.

(iv) Oxidation of the coating may be controlled by the use of an inert gas surrounding the torch impact work area.

(v) The ratio of electrical power to gas flow determines the heat content of the plasma stream, the velocity of the stream and the melting and deposition potential of the torch. Materials respond differently in the plasma. Gas flow rate and electrical power must be determined for each special application.

The factors that control the torch are electrical power input and stability, plasma forming gas, volume and velocity of carrier gases, configuration of torch electrodes, externally applied magnetic effects, heat content of the gas and efficiency of the torch cooling system.

Substrate cooling Because the plasma flame imparts a large amount of heat to the work piece, rapid dissipation of heat at the point of flame-work contact is necessary to prevent the overheating of deposits. Most substrate material should not be permitted to exceed 200°C in temperature during spraying. Substrate cooling is achieved by the internal circulation of on both sides of an external blast of inert gas, impinging on the substrate cold water or by the torch flame.

Powders for plasma spraying It is a matter of common knowledge that the quality of coating is materially affected by the powder quality; nevertheless, this problem has not been fully studied so far. Most of the authors use commercially available materials and only observe the grain sizes and their distribution in unit volume. The optimum grain size for a given spraying distance has been studied in Table 4.14.

Table 4.14

Typical Spray Coating data

Powder	Type	Size (M)	Substrate
ZrO_2	Fused, sintered chemically treated	74-104	304 Stainless steel aluminium zircoloy mild steel
UO_2	Fused	43-74	—do—
TiC	Fused	74-147	Beryllium
ZrC	High purity	74-147	—do—
TiN	Technical grade	74-147	—do—

In addition to the size of the particle, its shape also affects the transfer characteristics. An approximately spherical grain size is favourable because such powders are free flowing and can be conveniently converted and charged. Their feed is easily controlled and the quality of the coating layers is reproducible.

ELECTRON BEAM MACHINING (EBM)

Electron beams are now used in many industrial equipments. They have many special characteristics which make them most suited for specific applications. The most important of these characteristics is the high resolution and the long depth of field that is obtained because of the short wavelength of high energy electrons. Other features of the beams include their extraordinary energy (e.g. power densities of 10^6 kW/cm^2 have been achieved), ability to catalyze many chemical reactions, controllability and compatibility with high vacuum.

Electron beam machining (EBM) can be classified into two types. In one, which can be termed 'thermal type', the beam is used to heat the material up to the point where it is selectively vapourized. The other, called 'nonthermal type' utilizes the beam to cause a chemical reaction.

The ability of electron beams to cause drastic thermal effects has been known since a long time. In the late 1930s it was used for drilling the apertures of electron microscopes. In 1950, Steigerwald and his colleagues at Carl Zeiss A.G. (Germany) developed an electron beam milling machine. E.B. Bas, in 1960, made fine holes in ruby crystals. In attempting to drill holes in diamond, Steigerwald found that the electron beam could raise the temperature of diamond to as much as 3,000°C even though diamond tends to disintegrate into graphite at 2,000°C. By introducing oxygen into the vacuum system, CO or CO_2 could be formed at 300°C, thus burning away the material which was not possible to vaporize.

The ability of low energy electrons to bring about a surface chemical reaction was first observed and studied by P.H. Carr. Significant research in both thermal and nonthermal EBM has since then been done by many researchers.

A typical electron beam installation for drilling is shown in Fig. 4.34. It consists of five basic units: (i) electron gun which generates and directs a controlled beam of electrons of high energy density on the work material to change it chemically and physically, (ii) vacuum chamber and a high vacuum pump system, (iii) movable table within the chamber for mounting the work piece, (iv) an electronic system which controls the size and movement of the beam and (v) a monitoring instrument.

Generation and Control of Electron Beam

The electron beam is a stream of negatively charged particles which are generated, accelerated, and to some extent, focussed inside a device called an 'eletron gun'. The essential constituents of an electron gun are:

(i) A cathode, which serves as the source of electrons. It may be a current-carrying, self heating filament, or a solid block indirectly heated by radiation from a filament.

(ii) A grid cup, which is negatively biased with respect to the filament.

(iii) An anode which is kept at ground potential, and through which the high velocity electrons pass.

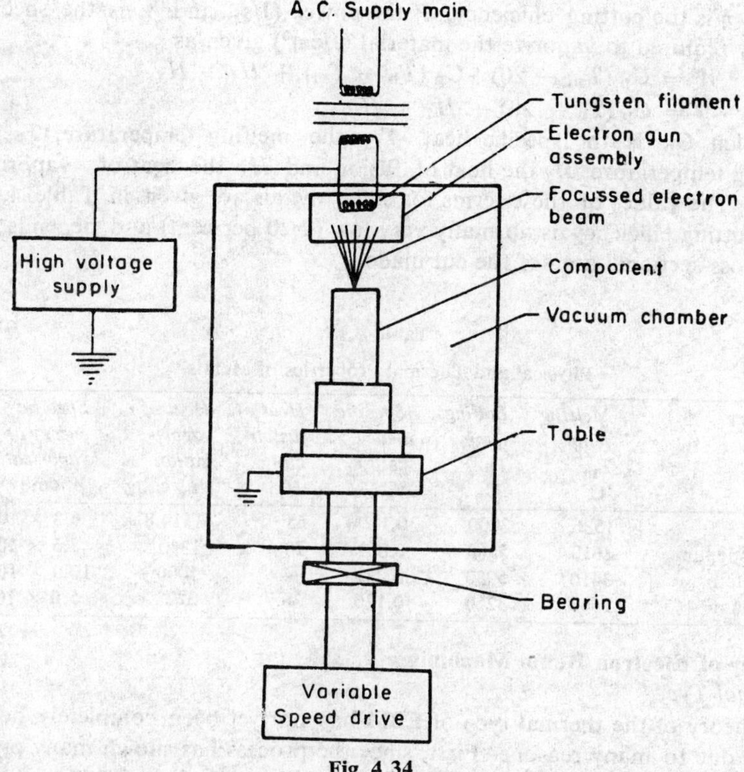

Fig. 4.34

The beam of electrons is emitted from the tip of the hot cathode. It is accelerated towards the anode by the high potential applied between the anode and the cathode. It passes through the anode at a speed up to two-third of the velocity of light. The flow of electrons is controlled by the negative bias applied to the grid cup.

A magnetic deflection coil fitted below the electron gun is used to make the electron beam circular in cross-section and deflect it anywhere. The beam is generated in a high vacuum for two reasons: (i) the emitter would rapidly oxidise when incandescent, at anything like atmospheric pressure, and (ii) the electrons would lose energy by collision with air molecules. The electron emitter with its focusing coils is used at a vacuum of 10^{-4} or 10^{-5} torr.

A power density of as high a magnitude as a billion Watts/square cm can be attained in the beam. This is sufficient to immediately fuse and vapourize any material on which it falls. Thus, in a thermal type EBM, cutting is, in fact, a precisely controlled vaporization process.

The rate R at which the material is vaporized can be calculated by

$$R = \eta \frac{P}{W} \tag{4.15}$$

where η is the cutting efficiency, P the power (J/s), and W is the specific energy required to vaporize the material (J/cm³) given as

$$W = C_p (T_m - 20) + C_p (T_b - T_m) + H_f + H_v$$
$$= C_p (T_b - 20) + H_f + H_v \qquad (4.16)$$

in which C_p is the specific heat, T_m the melting temperature, T_b the boiling temperature, H_f the heat of fusion and H_v the heat of vaporization. The values of these terms for some metals are given in Table 4.15. The cutting efficiency is normally very low (2-20 per cent) and depends on the cross-sectional area of the cut made.

Table 4.15

Physical and Thermal Properties of Metals

Property / Metal	Melting temp. T_m °C	Boiling temp. T_b °C	Specific heat C cal/g/ °C	Heat of fusion cal/g H_f	Heat of vaporization H_v cal/g	Specific energy of vaporization W/cm³
Iron	1536	3000	0.11	65	1514.8	6.3×10^4
Molybdenum	2610	5560	0.061	70	1340	7.5×10^4
Tungsten	3410	5930	0.032	44	1006	10.0×10^4
Titanium	1668	3260	0.126	36.7	2223	5.0×10^4

Theory of Electron Beam Machining

Thermal Type

The theory of the thermal type of EBM has not yet been completely developed due to many reasons. First, since the process has shown many practical results, no pressure has been put on the researchers in the field to develop extensive theories or to make difficult analyses. Second, although the individual physical processes involved are simple, their number, non-linearity and complicated geometry, all combine to make a theoretical analysis of the process difficult. Third, the small dimensions, short time intervals and lack of thermo-dynamic equilibrium make it difficult to get the required experimental data for correlation with any theory.

Forces in machining Since the temperatures in the vicinity of the electron beam cause the material to be in a molten state, the material is acted upon by forces of surface tension, gravity, electron beam pressure and the reaction force of the departing vapourized material. In machining non-conducting materials, electrostatic forces may also be present. A study of these forces (Fig. 4.35) will be made and the relationships established to find their magnitudes.

(a) *Electron pressure* The pressure due to electron bombardment can be estimated from the momentum loss suffered by the electron beam as it impinges upon the work surface.

$$f_e = m_e v_e \frac{i}{e}$$

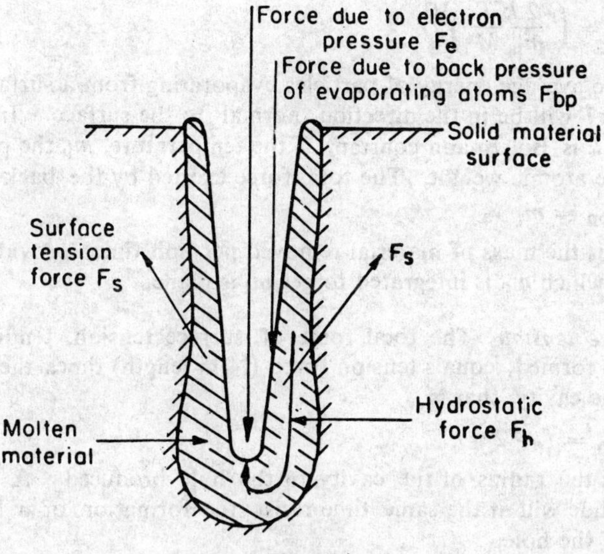

Fig. 4.35

where f_e is the electron pressure (N/m²), m_e the electronic mass, V_e the velocity (m/s), i the current density (amp/m²) and e the electron charge. The velocity v_e can be calculated from the accelerating voltage U as

$$v_e = \left(\frac{2\,e\,U}{m_e} \right)^{1/2} \tag{4.17}$$

Thus

$$f_e = \left(\frac{2\,m_e U}{e} \right)^{1/2} i \tag{4.18}$$

The total force due to election bombardment is

$$F_e = \pi\, r^2 f_e$$
$$= \pi\, r^2 \left(\frac{2\,m_e\,U}{e} \right)^{1/2} i$$
$$= \left(\frac{2\,m_e\,U\,I^2}{e} \right)^{1/2} \tag{4.19}$$

where r is the radius of the hole and I is the total current.

(b) *Back pressure of evaporating atoms*   The effect of momentum of the thermally evaporating material can be studied by considering the conservation of momentum

$$p_a = m_0\,v_a \tag{4.20}$$

where p_a is the back pressure of the evaporating atoms, m_0 the mass of material removed per unit area of surface per unit time, v_a the atomic velocity given as

$$v_a = \left(\frac{2\,k\,T}{m_p\,M} \right)^{1/2} \tag{4.21}$$

because the average energy of particles evaporating from a surface is $2\,kT$, of which kT will be in the direction normal to the surface. In the above equation, k is Boltzmann constant, T the temperature, m_p the proton mass and M the atomic weight. The total force exerted by the back pressure is

$$F_{bp} = m_1\,v_a \tag{4.22}$$

where m_1 is the mass of material removed per unit time, provided the surface over which m_o is integrated to get m_1 is plane.

(c) *Surface tension* The total force of surface tension, tending to close the cavity formed, equals tension force (N/m length) times the circumference of the cavity, that is

$$F_s = f_s \cdot 2\,\pi\,r \tag{4.23}$$

where r is the radius of the cavity or the hole produced. A force equal in magnitude will at the same time resist the formation of a liquid lip at the top of the hole.

Surface tension has not been studied till now for liquid metals at such high temperatures as are involved in EBM. However, the values are available for many metals near their melting points. For gold, silver, copper, etc., the values are in the range of 500-2,000 dynes/cm.

(d) *Hydrostatic pressure of molten metal* The hydrostatic pressure due to the molten surface on the side of the cavity being generated is

$$f_h = \rho\,g\,h \tag{4.24}$$

and the total force exerted on the bottom of the hole towards the top is

$$F_h = \rho\,g\,h\,\pi\,r^2 \tag{4.25}$$

where ρ is the density of the molten metal (kg/m³), g the acceleration due to gravity, h the depth of the cavity, and r the radius of the cavity formed.

In the calculation with actual data, the hydrostatic force is normally very small and can be neglected. The electron pressure force is the largest of all these forces but the other two forces are also significant. If the surface tension force is larger than the atomic reaction force, molten metal would tend to flow from the walls into the bottom of the wall, where it would be evaporated by the beam; otherwise, if the reaction force is larger than the surface tension force (case of big holes), molten material would be pushed out of the hole.

Nonthermal Type

Christy, who has made an excellent study in this field, has put forward a phenomenological theory which predicts the rate of film growth for the polymerization of organic films by electron bombardment. His theory is summarized here.

Consider a surface on which a film is being deposited as a result of the polymerization of molecules arriving at rate n_0 per unit area per unit time by the action of electrons arriving at rate n_e per unit area per unit time. Further, let a denote the cross-section for reaction, τ the mean time of stay of an unreacted molecule on the surface, N_0 the number of unreacted molecules, and N_1 the number of reacted particles on the surface. It can be assumed that a reacted molecule remains on the surface permanently. Also, it is assumed that the initial surface is made up of already polymerized molecules so that no anomalous surface effect takes place for the first layer, then

$$\frac{dN_1}{dt} = a\, n_e\, N_0 \tag{4.26}$$

and

$$\frac{d N_0}{dt} = n_0 - \frac{N_0}{\tau} - \frac{d N_1}{dt} \tag{4.27}$$

By solving these equations for N_0

$$N_0 = \frac{n_0}{a\, n_e + \dfrac{1}{\tau}}\left\{ 1 - K\left[\exp - \left(a\, n_e + \frac{1}{\tau}\right) t\, \right]\right\} \tag{4.28}$$

where K is a dimensionless constant and depends on the initial surface coverage. $K = 0$ corresponds to the steady state solution; $K \leqslant 0$ or $K > 0$ to an initial concentration greater or less, respectively, than the steady state condition. The following cases arise.

Case 1 This case corresponds to the condition when surface density of unreacted molecules N_0 is less than that of one monolayer. This means that $N_0 < \dfrac{1}{a_0}$, where a_0 is the surface area of one molecule. Thus, from Eq. 4.28 we have

$$n_0 < \left(\frac{a}{a_0}\right) n_e + \frac{1}{a_0\, \tau} \tag{4.29}$$

Also from Eqs. 4.26 and 4.28

$$\frac{d N_1}{dt} = \frac{n_0}{1 + \dfrac{1}{a n_e \tau}}\left\{ 1 - K\left[\exp - \left(a\, n_e + \frac{1}{\tau}\right) t\, \right]\right\} \tag{4.30}$$

and on integration

$$N_1 = \frac{n_0}{1 + \dfrac{1}{a n_e\, \tau}}\left[t + \frac{K}{a\, n_e + \dfrac{1}{\tau}} \exp - \left(a\, n_e + \frac{1}{\tau}\right) t\, \right] \tag{4.31}$$

It is assumed that $N_1 = 0$ at $t = 0$. For $K = 1$ corresponding to the initial coverage and $t \ll \left(a\, n_e + \dfrac{1}{\tau}\right)$, $\dfrac{d N_1}{dt}$ is proportional to t, and N_1 is proportional to t^2. In steady state conditions, the rate of film growth can be written as

$$R_t = \frac{d\,N_1}{dt} \cdot V_m \tag{4.32}$$

where V_m is the volume of one molecule.   Thus

$$R_t = \frac{n_0\,V_m}{1 + \dfrac{1}{a\,n_e\,\tau}} \tag{4.33}$$

Case 2   This is the case when the arrival rate of molecules is sufficient to maintain a monolayer on the surface at all times.   The case is restricted to the situation $N_0 = \dfrac{1}{a_0}$, that is, more than one monolayer is not possible because of a significant lowering of the binding energy of molecules in the second layer.   The condition can be visualized by considering that T approaches zero for the second layer, thus

$$n_0 \geqslant \left(\frac{a}{a_0}\right) n_e + \frac{1}{a_0\,\tau} \tag{4.34}$$

$$N_0 = \frac{1}{a_0} \tag{4.35}$$

and

$$R_t = \left(\frac{a}{a_0}\right) n_e\,V_m \tag{4.36}$$

Case 3   This case corresponds to the condition when all the arriving molecules stick to the surface regardless of the surface conditions, that is, τ approaches infinity and a thick film of unreacted material is formed. Equations (4.28), (4.29) and (4.31) transform to

$$N_0 = \frac{n_0}{a\,n_e} + K' \exp\left(-\,a\,n_e\,t\right) \tag{4.37}$$

$$\frac{d\,N_1}{dt} = n_0 + a\,n_e\,K' \exp\left(-\,a\,n_e\,t\right) \tag{4.38}$$

$$N_1 = n_0\,t - K'\left[\exp\left(-\,a\,n_e\,t\right) - 1\right] \tag{4.39}$$

where K' is another dimensionless constant and it has been assumed that the probability of a given molecule reacting is not dependent upon its depth below the surface.   This means that the total film is quite thin and is transparent to the bombarding electrons.   The rate of deposition of reacted molecules under steady state conditions is given by

$$R_t = n_0\,V_m \tag{4.40}$$

Case 4   This case refers to the situation when the unreacted film is completely deposited before the electron bombardment. In this case, we will have $x = 0$ and

$$N_0 = K' \exp(-a\, n_e\, t) \qquad\qquad (4.41)$$

$$\frac{dN_1}{dt} = a\, n_e\, K' \exp(-a\, n_e\, t) \qquad\qquad (4.42)$$

$$N_1 = K'\,[\,1 - \exp(-a\, n_e\, t)\,] \qquad\qquad (4.43)$$

The deposition rate calculations in this case have no meaning. It can be seen, however, that to form a film K' thick, containing less than 1 per cent of unreacted molecules, bombardment is required for a time of $\dfrac{5}{a\, n_e}$ because $e^{-5} = \dfrac{1}{150}$.

Christy has also experimentally studied the formation of polymerized organic films. He has shown that the product $a\, n_e$ depends exponentially upon the reciprocal temperature due to the temperature dependence of τ. Ennos also studied this phenomenon. He observed that reaction cross-section remains almost constant in the electron energy range of 2-74 k ev.

Process Capabilities and Limitations

The main uses of this process today include the cutting and welding of materials. The chief advantage of cutting from the engineer's point of view is that the process is not dependent on the work material properties. The materials which can be cut include aluminium, beryllium, cemented carbides, ceramics, copper alloys, glass, alloy steels, tantalum, titanium, tungsten and zirconium. The process works as effectively on extremely hard and tough alloys as on soft nonferrous metals.

Though the process does not possess the advantage of high material removal rates, it is claimed that it can be used to cut very accurate slots and shapes in all kinds of materials.

The electron beam machining has a promising future in the cutting of delicate and consistent shapes which are needed in certain electronic assemblies. In such applications, it can be a formidable competitor of ultrasonic and chemical machining processes which are currently used for this type of work.

Comparison of Thermal and Nonthermal Processes

The two processes are not as much in competition as they initially may appear to be. For drilling holes and making slots or other deep constructions, the thermal process is better because the high energy density required necessitates working on a small area only. Due to the fact that electrons give up most of their energy at approximately 5 to 15 microns below the surface, the work surface cannot be vapourized without melting the material in this region, and therefore, the use of this process for machining the top layer of a thick laminar structure is not recommended. The thermal process has reached the technical stage where it can find immediate industrial use but the nonthermal process has not.

The nonthermal process is particularly useful for machining large areas of thin films. Greater depths cannot be attained due to very low reaction rates but resolution is comparatively much superior. The material removal rate per unit area in this process is only 10-20 per cent of that achievable in the thermal process.

NEUTRAL PARTICLE ETCHING

Although electron beam machining is not related directly to this process, It is briefly described here, as a related topic.

The process entails a neutral particle arriving at a surface, meeting with a surface atom, and the resulting molecule evaporating. The rate of material etched is a function of the arrival rate of etching particles. Figure 4.36 shows the geometry factors involved in the process. The bombarding particles density N_0 can be calculated as

$$N_0 = \frac{p}{(2\pi \, M \, m_p \, k \, T)^{1/2}} \left(\frac{h'}{2\,l} \right)^2 \tag{4.44}$$

where p and T are respectively the pressure and temperature of the etchant at the source, M is the atomic weight of the etching atom, m_p is the proton mass, k is Boltzmann constant and h' and l are as given in Fig. 4.36. For drilling a hole with a large depth to diameter ratio, the beam is adjusted to approximately the same ratio, that is, small h/l. The pressure must also be low enough so that the beam adjustment does not cause the collision of particles in the beam.

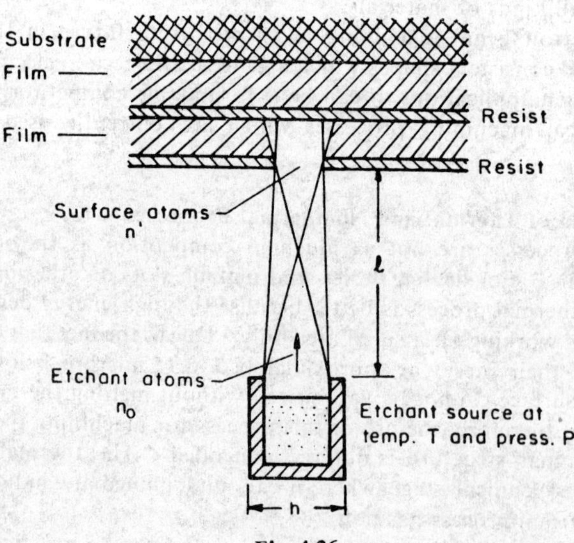

Fig. 4.36

Let x be the number of bombarding atoms when combined take away y number of surface atoms by chemical reaction

$$x\theta + y\psi \to \theta_x \psi_y \tag{4.45}$$

where θ and ψ are respectively atomic species in the beam and at the surface.

The number of surface atoms n' leaving the surface per unit area per unit time is calculated as

$$n' = (y/x) \, n_0 \tag{4.46}$$

In case a fraction of the arriving atoms leaves the surface without reacting, Eq. 4.46 will have to be modified. However, this should be avoided as far as possible; otherwise, if atoms, on reaching the bottom of a hole are machined, they will run unreacted to the sides of the hole and cause under-cutting. Not taking this into account in our calculations, the machining rate is given by

$$m_0 = M \, m_p \, \frac{y}{x} \, n_0$$

$$= p \left(\frac{M \, m_p}{2 \pi k T} \right)^{1/2} \frac{y}{x} \left(\frac{h}{2l} \right)^2 \tag{4.47}$$

Example

A molybdenum surface has to be machined by chlorine atoms to form molybdenum trichloride. To get nascent chlorine, a generator is provided. The value of $p = 0.13$ N/m². Assuming the temperature of etchant to be 300 K, estimate the machining rate.

Solution

Here

$$M = 35 \text{ (for chlorine)}$$
$$y/x = 1/3$$
$$h/l = 1/5$$

Using Eq. 4.47, the machining rate can be found to be 2.7×10^{-6} kg/m²/s.

LASER BEAM MACHINING (LBM)

The word LASER stands for 'Light Amplification using Stimulated Emission of Radiation'.

Lasers provide intense and uni-directional beams of light. This light is coherent in nature, whereas ordinary light is particularly incoherent because in the latter case, the source of light is generally a hot matter.

The laser in short pulses has a power output of nearly 10 kW/cm² of the beam cross-section. (The sun's total radiation is 7 kW/cm² of its surface.)

By focussing a laser beam on a spot 1/100 of a square mm in size, the beam can be concentrated in a short flash to a power density of 100,000 kW/cm². This can provide enough heat to melt and vapourize any known high strength engineering material and permit the fusion and welding of refractory materials.

In a laboratory test when a laser beam was focussed on a piece of carbon, a spot was heated to approx 8,000 K in 0.0005 second.

Thus, it is seen that the laser promises to be a valuable tool for drilling, cutting or milling virtually any metal, ceramic or other organic material. Any material that can be melted without decomposition can be cut with the laser beam.

Apparatus

The most important part of the laser apparatus is the laser crystal. Many materials with laser action have been developed, for example, calcium fluoride crystals doped with neodymium ($Ca+F_2+Nd$) and glass doped with various rare earths. The most commonly used laser crystal is a man-made ruby consisting of aluminium oxide into which 0.05 per cent chromium has been introduced ($Al_2O_3+Cr_2$).

The crystal rods are usually round and the end surfaces are made reflective. A laser rod for a 3 Joule unit is about 6 mm in diameter and 70 mm in length.

The laser rod is excited by the xenon filled flash lamp which surrounds it. Both are enclosed in a highly reflective cylinder which directs the light from the flash lamp into the rod. The chromium atoms in the ruby are thus excited to high energy levels. The excited ions emit energy (photons) when they return to the normal state. In this way, very high energy is obtained in short pulses. The ruby rod becomes less efficient at higher temperatures; it is thus continuously cooled with water, air or liquid nitrogen.

Material Removal

The mechanism by which a laser beam removes material from the surface being worked involves a combination of the melting and evaporation processes. However, with some materials, the mechanism is purely one of evaporation.

Laser machining is essentially a thermal process; the heat requirements and the heat utilization involved are discussed below.

Thermal Features of Laser Machining

The radiant energy delivered to a surface by a focussed laser beam is consumed in four ways:

(i) A part is reflected and lost (this part becomes larger in the case of highly reflective metal surfaces.

(ii) Most of the energy which is not reflected is used for melting metal.

(iii) A relatively small part of the energy is used to evaporate the liquid metal.

(iv) A very small part of the energy is conducted into the unmelted base material.

The relative magnitudes of the heat consumption at these four avenues strongly depend upon the thermal and optical properties of the material being worked and the intensity and pulse duration of the laser beam.

It is seen that all the energy that leaves the laser head does not reach the material surface; some is absorbed on the way. Part of the material expelled from the surface stays in the path of the beam in the form of small droplets and continues to absorb energy. This represents a definite loss in thermal efficiency, the importance of which has not yet been quantitatively assessed.

Thermal Analysis

A simple case will be analysed in which the intensity of the laser beam does not vary with time during the pulse and is uniform over the entire area of the hot spot.

Machining by laser is a high speed ablation process. As a uniform high intensity energy is delivered to a work surface, there is an initial transient period, after which steady state ablation is established, and this continues as long as the energy continues to be supplied at uniform intensity. This steady state ablation is characterized by a constant rate of material removal (weight/unit area and time) and by the establishment of a steady temperature distribution in the solid immediately in advance of the ablating surface.

In cases where the material being worked is both melting and evaporating, the steady temperature distribution is given by

$$\frac{T - T_1}{T_m - T_1} = \exp\left\{\frac{-V_{ab}\,x}{\beta}\right\} \tag{4.48}$$

where

T—temperature at distance x below the ablating surface,

T_1—initial temperature (uniform) of the material being worked,

—temperature at a large distance from the ablating surface,

T_m—melting point of the material being worked,

v_{ab}—steady ablation velocity,

$\beta = \dfrac{s}{\rho C_p}$— thermal diffusivity of material being worked,

$s,\ \rho,\ C_p$ — thermal conductivity, density, and specific heat, respectively of the material being worked.

Since x is measured from a moving boundary, that is, the ablating surface, the temperature distribution is in terms of a moving coordinate system.

It can be seen that the exponential temperature distribution represented by the above equation satisfies the boundary conditions that $T = T_1$ when x is very large, and $T = T_m$ when $x = 0$. The depth up to which heat penetrates beyond the ablating surface is of considerable practical importance, and it is required to be as shallow as possible.

The term 'characteristic depth' x_c is defined as the depth during steady ablation which has experienced a temperature rise of $1/2.718$ times $(T_m - T_1)$ from the temperature T_1. This x_c decreases with increasing ablation velocity and decreasing thermal diffusivity.

$$x_c = \frac{\beta}{v_{ab}} \tag{4.49}$$

During the initial transient period, when ablation is just beginning, part of the heat delivered to the work surface is used to establish the temperature distribution within the solid. Once steady conditions are obtained, the heat contained in the solid does not increase any further and the value of this steady heat content is given by

$$\left(\frac{H}{A}\right)_0 = \int_0^\infty \rho \cdot C_p \, (T - T_1) \, dx$$

$$= \frac{s \, (T_m - T_1)}{v_{ab}} \tag{4.50}$$

where $(H/A)_0$ is the heat per unit area contained in the solid beneath the ablating surface.

After the heat contained in the solid has reached its steady value, all the heat at the material surface (less the amount reflected) is used for melting and vapourizing material. The relationship governing it is

$$f' = v_{ab} \, \rho \, H \quad \text{or,} \quad v_{ab} = \frac{f'}{\rho H} \tag{4.51}$$

where

f' — net heat flow rate per unit area (total−reflected)

H — heat requirement for melting and vapourization or the heat/unit weight of material required to elevate the temperature of the solid from its initial value to the melting point, plus the heat of fusion and the additional heat supplied to boil a small fraction of the metal which has been melted.

The two parameters f' and H are, however, most difficult to assess quantitatively. The first requires knowledge of not only the beam intensity but the average effective reflectivity of the material during the pulse; the second requires a good estimate of exactly how much metal is vapourized and how much remains in the liquid state.

Cutting Speed and Accuracy of Cut

The latest commercially available laser unit of 800 W capacity can cut most metal plates up to 3 mm thick at speeds of about 1 to 1.25 m/min.

The general cutting accuracy of this unit is claimed to be within 0.8 mm. A feature of the cutting process is that a square and straight edge is obtained.

Metallurgical Effects
The heat affected zone is relatively narrow, about 1 mm for 3 mm thick sheet metal. It is reported that at these thin sheets, very little melting occurs and there is almost instantaneous vapourization. Therefore, there is no significant effect on metallurgical properties of work material.

Advantages
Any soild material which can be melted without decompositon can be cut with the laser beam. The other major advantages of the laser beam as a cutting tool are:
  (i) There is no mechanical contact between the tool and the work.
  (ii) Large mechanical forces are not exerted upon the work piece.
  (iii) The laser operates in any transparent environment, including air, inert gas, vacuum and even certain liquids.
  (iv) The laser head need not be in close proximity for performing cutting and drilling operations in locations of difficult accessibility.
  (v) The beam can be projected through a transparent window.
  (vi) Unlike other thermal machining devices, the laser can be used with materials sensitive to heat shock such as ceramics. (This advantage is also obtained with electron beam.)
  All features of laser beam machining improve with increased intensity. The higher the intensity, the lesser is the heat resonant in the uncut material.

Limitations
One main limitation of the laser cutting is that currently it cannot be used to cut metals that have high heat conductivity or high reflectivity. This means that it cannot be used to cut aluminium, copper and their alloys satisfactorily. This limitation, although not very important, can even be regarded as an advantage, in that the laser beam is a self-selective cutter for different materials. This will permit the metal that is to be cut to be placed on a work table made from another metal which is not affected by the laser beam, and thus give complete manoeuvrability in cutting out complex profiles without damage to the work table.
  During operation, the work piece to be cut is placed on the aluminium work table (which is resistant to being cut by laser beam). The laser head is traversed over the work piece and an operator visually inspects the cut while manually adjusting the control panel. The actual profile is obtained from a linked mechanism which copies the master drawing or actual profile placed on a nearby bench.
  The other limitations of this method are:
  (i) The machined area can be irregular due to off-axis modes that may be generated during laser action.
  (ii) The least diameter to which laser beam can be focussed depends upon the laser beam divergence; this in turn, is a function of the quality of the laser material and the laser cavity length.
  (iii) Output energy from laser is difficult to control precisely.
  (iv) The laser system is quite inefficient.

(v) Pulse repetition rates are low.

Laser Machine

Extensive research in the development of efficient and high power laser units was carried out in the last decade. In 1967, a typical 500 W laser unit was about 19 m long. By 1969, a 600 W device was developed which measured only 4 m in length. In 1970, Ferranti Professional Component Department (an English firm) developed 800 W laser unit, only $1\frac{1}{2}$ m long.

HOT MACHINING

The process of hot machining was developed to overcome the problems of low cutting speeds and feeds and heavier load on the machine bearings and slides which are commonly encountered when machining newer, hard and tough alloys.

The process of hot machining is simple and convenient in the sense that it can make use of the conventional machine tool. It utilizes the heat applied locally by some external source just ahead of the cutting tool to reduce the shear strength of the component material. This permits easy formation of the cutting chip accomplished with lessened shocks to the tool and comparatively good surface finish on the work material. There is little evidence of any adverse effect on the microstructure of the work material.

The amount and place of application of heat is of great importance in this process. The heating requirements and the methods available to heat the workpiece are discussed below.

Heating Requirements

The heating method should be so designed as to satisfy the following requirements:

(i) It should provide a high rate of heat input so that the work material is raised to the specified temperature in a short time.

(ii) It should allow heating of the shear zone only, as the penetration of heat to too great a depth can cause thermal damage.

(iii) It should provide a wide range of constant temperatures.

(iv) It should be easy to set up and control.

(v) The capital investment as well as operating cost should be low.

Methods of Heating

Several methods of heating to cover the various machining applications are available. The selection of a specific method depends not only on the component material but also on the shape or type of the component. Some of the heating methods which have found use in certain applications are discussed below.

Flame Heating

In this method, a gas flame is employed to heat the component material just ahead of the cutting tool. Either oxy-acetylene, propane or town gas can be used. In order to increase the heat input and its concentration, a multi-jet head may be used.

It is a versatile method and uses relatively inexpensive equipment. It has been found to be very economical for milling long but narrow jobs. For wider jobs, the problems of the localization of heat and the maintenance of a constant temperature over a larger area make this method unsuitable. Further, there is a risk of overheating or oxidation of work material which necessitates some post-hot machining operation, like grinding, in some cases.

D.C. Arc Heating

In this method, an electric arc produced with carbon (or tungsten) electrode and the work material is used to heat nonmagnetic or refractory materials required to be hot machined. The negative terminal of the arc welding machine is connected to the work piece and the positive terminal to the electrode which draws the heating arc just ahead of the cutting tool. A magnetic field may be used to direct and to prevent the arc from wandering.

The maintenance of a constant temperature of work material is, however, a problem with this method.

Plasma Arc Heating

The plasma arc employs a restricted arc through which a selected gas is made to flow. This method has been satisfactorily used for heating most of the low permeable materials since temperatures can easily be controlled. The process can provide temperatures as high as 20,000 K and heating rates can be controlled to prevent surface damage. This is significant because metals of low impedence are usually relatively insensitive to heat damage.

The initial equipment cost is high but operating costs are low. Large areas of metallic work pieces can be heated and machined at high rates if provision for adequate power and gas supplies is made.

To avoid the effects of radiation the same safety precautions, as for 'Plasma Arc Machining', must be observed.

Furnace Heating

This is one of the simplest methods of heating the workpiece. A gas or electrical furnace is used, the temperature of which is raised to the required degree for the machining operation. The work material is placed in the furnace and allowed to remain there until its surface is at the same temperature as that of the furnace. It is then taken out and held in a fixture on the machine on which it is to be machined. No time should be lost between the removal of the work material from the furnance and the start of the machining operation.

This method of heating is economical when a furnance of the required type already exists and the metal cutting machine tool can be located adjacent to the furnance. It is difficult to control the depth of the heating zone, and thus, there exists the possibility of workpiece damage due to uneven cooling.

Resistance Heating

This is another simple method of heating the work piece which involves passing 50 cycle A.C. through the component part or through the heating elements incorporated in the body of the work holding fixture.

As the component temperature can be kept fairly constant without much difficulty, there is little adverse effect of distortion and dimensional discrepancy caused by uneven cooling. One main precaution to be observed in this process is that arcing must not be allowed to occur.

This method of heating is simple and relatively inexpensive. It is particularly suitable for the milling of small area jobs.

Radio Frequency Resistance Heating

This method of heating makes use of radio frequency currents which follow the path of least impedence. The equipment used in this method permits the transfer of current to the workpiece and a low impedence path is provided across it so that the applied current heats only the predetermined area. The required low impedence path is obtained by placing a conductor (called return conductor) very close to the workpiece so that the current is made to flow in a direction opposite to the current path in the component.

The main advantage of this method is that heat is localized in the surface area as well as in the depth with high input.

This method is gaining popularity and it appears that further advances in the design of equipment for this method will definitely boost the application of hot machining techniques. It has presently been applied successfully for end and face milling and for turning operations. Arcing must be prevented in this method as in conventional resistance heating methods.

Induction Heating

This method of heating makes use of an alternating current at a frequency of 3,000 c/s to 1.2 Mc/s. The heating current is induced in the component by transformer action with the help of a primary coil which is designed to cover the area to be machined.

This method also provides localized heat without much difficulty. The control of temperatue, required depth of heat zone and repeatability of conditions can be achieved without undue problems.

The equipment required for this method of heating is relatively costly but it has many other applications within an engineering works. Sometimes, difficulty may be faced in mounting a workpiece on the machine. This may be overcome in many cases by the use of heating coils of the required size and shape.

Tool Life and Production Rate

An increase in tool life is possible when the component which is to be machined is heated to about 800 K. Higher temperatures connot be employed due to the problem of welding of chip to the tool. In case of face milling, it has been reported that the use of a slab of asbestos at the entry side of the component can lead to increased tool life. It is felt that the asbestos tends to act as a damper to reduce the severity of the impact shock between the teeth and the component. A chemical reaction is assumed to be set up by the heated impact which serves a lubricating function.

Cutting speeds up to 800 rpm have been achieved on lathe and milling machines with metal removal rates up to 30 cm^3/min/kW. After selecting the most suitable heating temperature to suit the material of the component, the machining is done as fast as possible, commensurate with the required surface finish. If the component is subjected to high temperatures for too long a period, it will temper and lose its hardness, which may be undesirable. Apart from the possible adverse effects of tempering, there has been, so far, no evidence of any important change in microstructure.

REVIEW QUESTIONS

1. In an EDM operation employing relaxation circuit, discuss the effects of (i) charging resistance, (ii) gap setting and (iii) capacitance on the rate of metal removal.
   How does this type of machine compare with the pulse generator?
2. In an EDM with R-C circuit, the supply voltage is 100 V and the break-down voltage corresponds to maximum power delivery conditions. If the supply voltage is increased to 125 V without altering any of the other electrical parameters of the circuit, what is the percentage increase in metal removal rate that can be expected?
3. In an EDM operation, with R-C circuit, the following data are available.

Supply voltage	100 V
Discharge voltage	75 V
Resistance (R)	10 Ω
Percentage of discharge energy used up in metal removal operation	20 per cent

Calculate the time required to drill a 10 mm diameter hole in a steel work piece, having a thickness of 15 mm.

4. Discuss the factors influencing the choice of electrode material in EDM. Name the best electrode material for finish machining a small die made of WC by the EDM process.

5. Assuming that sparking in EDM leads to the formation of spherical craters on the work surface, calculate the centre line average value of the surface obtained during the process when the circuit capacitance is 1.0 μF, the discharge occurs at 50 V and 45 per cent of the spark energy is used up in metal removal. The dielectric used is kerosene. It the supply voltage is increased to 60 V and the capacitance is reduced to 0.8 μF, what is the per cent change in the surface finish value that would be obtained?

6. Discuss the advantages of EDM as compared to other nontraditional methods with regard to (i) metal removal rate, (ii) accuracy, and (iii) surface finish.

7. Explain what is meant by nontransferred and transferred mode of plasma arc. What are the advantages of each?

8. Discuss some of the important considerations in the design of a plasma torch. What are the essential differences between **a** cutting and a welding torch?

9. Discuss the factors that influence the quality of the cut in PAM.

10. What are the essential differences between plasma spray and oxy-gas spraying? Discuss in detail.

11. Describe, with the help of a sketch, the constructional features of an 'electron gun' used for generating an electron beam in electron beam machining.

12. Write short notes on
    (i) Process capabilities of electron beam machining.
    (ii) Comparison between thermal and nonthermal features of electron beam machining.

13. What is laser and how is it used to machine the materials? Give the thermal features and analysis of the laser beam machining.

14. Make a comparison between laser beam and electron beam machining processes on the basis of their applications and limitations.

15. (i) What is the method of hot machining?
    (ii) What are the possible harmful effects of this method on the work material?
    (iii) Write a short note on the requirements of a good heating method used in hot machining.

16. Describe any three methods of heating the work material in the process of hot machining. Give their relative advantages and disadvantages.

17. Sketch a set-up of hot machining process by induction heating. What are the chief advantages of employing this method of heating over furnance heating?

Bibliography

General

1. Merchant, M.E., "Recent progress in metal removal research at Cincinnati", *Int. J.M.T·D.R.*, No. 1/2, Vol. 1/2, 1961.
2. Springborn, R.K., "Nontraditional machining processes", *ASTME*, Michigan, 1967.

Ultrasonic Machining

3. Miller, G.E., "Special theory of ultrasonic machining", *Journal of Applied Physics*, No. 2, Vol. 28, p. 149, 1957.
4. Shaw, M.C., "Ultrasonic grinding", *Microtecnic*, Vol. 10, p. 257, 1956.
5. Rosenberg, L.D., V.F. Kazantsev *et al.*, "Ultrasonic Cutting", Authorized translation from Russian into English by J.E.S. Bradley (New York Consultants Bureau), 1964.
6. Kazantsev, V.F. and L.D. Rosenberg, "The mechanism of ultrasonic cutting", *Ultrasonics*, Vol. 3, p. 166, 1965.
7. Shreiner, L.A., *The Hardness of Brittle Bodies* (in Russian) (USSR), Academy of Sciences Publishing House, 1949.
8. Koifman, M.I., "The strength of hard particles", *Doklady Academi i Nauk*, Vol. 29, p. 477, 1940.
9. Goetze, D., "Effect of vibration amplitude, frequency and composition of abrasive slurry on the rate of ultrasonic machining in ketos tool steel", *Journal of Acoustic Society of America*, Vol. 28, p. 1033, 1956.
10. Pentland, W. and J.A. Ektermanis, "Improving ultrasonic machining rates, some feasibility studies", *Trans. ASME*, Series B, No. 1, Vol. 87, p. 46, 1965.
11. Neppiras, E.A. and R.D. Foskett, "Ultrasonic machining", *Phillips Technical Review*, Vol. 18, pp. 325, 368, 1956-57.
12. Markov, A.I., *Ultrasonic Machining of Intractable Materials*, Illife Books Limited, London, 1966.
13. Neppiras, E.A., "Report on ultrasonic machining" *Metalworking Production*, Vol. 100, pp. 1283-88, 1333-36, 1377-82, 1420-24, 1464, 1554-60, 1599-1604, 1956.
14. Metalkin, I.V. *et al.*, "Mechanical machining of various materials with ultrasonics", *Stanki Instruments*, No. 2, Vol. 16, 1956.
15. Neppiras, E.A., *Design of Ultrasonic Machine Tools*, Institution of Mechanical Engineers, Conference on Technology of Engineering Manufacture, March 1958.

16. Sheldon, G.L. and I. Finnie, "The mechanics of material removal in the erosive cutting of brittle materials", *Trans. ASME*, Series B, Vol. 88, p. 393, 1966.

17. Kennedy, D.K. and R.J. Grieve, "Ultrasonic machining—A review", *The Production Engineer*, Vol. 54, p. 481, September 1975.

18. Lvoie, F.J., "Abrasive jet machining", *Machine Design*, p. 135-39, September 1973.

19. Ingulli, C.N., "Abrasive jet machining", *Tool and Manufacturing Engineers*, No. 5, Vol. 59, p. 28, 1967.

20. Sheldon, G.L. and I. Finnie, "On ductile behaviour of normally brittle materials during erosive cutting", *Trans. ASME Journal of Engineering for Industry*, Vol. 88, p. 387, November 1966.

21. Bhattacharyya, A., *New Technology*, The Inst. of Engineers (India), 1973.

22. Sarkar, P.K. and P.C. Pandey, *Some Investigations on the Abrasive Jet Machining*, Paper presented at the Semi-Annual Meeting, Mech. Engg. Division, Inst. of Engineers (India), August 1975.

23. Anon., *Abrasive Machining*, *ASTME*, Collected Papers, Book 6, 1963.

24. Bitter, J.G.A., "A study of erosion phenomenon", *Wear*, Vol. 6, p. 5, 1963.

25. Roderts, A.G., W.A. Crouse and R.S. Pizer, "Abrasive jet method for measuring abrasion resistance of organic coatings", *ASTME* Bulletin No. 208, September 1955.

26. Pandey, P.C. and M.L. Neema, "Erosion of glass when acted upon by an abrasive jet", *Proc. Int. Conference on Wear*, St. Louis, p. 387, April 1977.

27 Sapra, I.L., "Studies on abrasive jet machining", *M.E. Dissertation*, University of Roorkee, 1975.

Water Jet Machining

28. Farmer, I.W. and P.B. Attewell, "Rock penetration by high velocity water jets", *International Journal of Rock Mechanics and Mineral Science*, No. 2, Vol. 2, p. 135, July 1965.

29. Brook, N. and D.A. Summers, "The penetration of rock by high-speed water jets", *International Journal of Rock Mechanics and Mineral Science*, No. 3, Vol. 6, p. 249, 1969.

30. Cooley, W.E. and L.L. Clipp, "High pressure water jets for undersea rock excavation", *Journal of Engineering for Industry, Trans. ASME*, Series B, Vol. 92, p. 281, May, 1970.

31. Beutin, E.F. *et al.*, "Grundlagen in der anwendung von diskontinuierlichen fluessig keitsstrahlen hoher geschwindigkeit", (Principles in the application of discontinuous high velocity jet fluids), *Werkstatt and Betrieb*, No. 6, Vol. 110, p. 369, June 1977.

32a. Franz, N.C., *Fluid Additives for Improving High Velocity Jet Cutting*, A 7-93, First International Symposium of Jet Cutting Technology, England, April 1972.

32b. Franz, N.C., "High energy liquid jet slitting of corrugated board", *TAPPI*, No. 6, Vol. 53, p. 111, June 1970.

33. Bryan, E.L., "High energy liquid jets as a new concept for wood machining", *Forest Production Journal*, No. 8, Vol. 13, p. 305, August, 1963.

34. Imanaka, O., S. Fugino, K. Shinohara and Y. Kawati, *Experimental Study of Machining Characteristics by Liquid Jets of High Power Density up to $10^8 W\ cm^{-2}$* G3-25, First International Symposium of Jet Cutting Technology, *BHRA* Fluid Engineering, Cranfield, England, April 1972.

35. Neusen, K.F. and E.C. Labrush, "Material removal by high pressure liquid jets at ten kilobars", *Journal of Engineering for Industry, Trans. ASME*, Series B, Vol. 97, p. 1067, August 1975.

36. Imanaka, O., "Machining with continuous liquid jets at pressures to 10 kbars", *Proc. International Conference on Production Engineering*, Tokyo, pt. 1, p. 26, 1974.

Electrochemical Machining

37. Opitz, H., "Electrical machining processes", *International Research in Production Engineering*, ASME, New York, p. 225, 1963.

38. Tipton, H., "The dynamics of electrochemical machining", *Advances in Machine Tool Design and Research*, Oxford, Pergamon Press, Ltd., p. 500, 1964.

39. Koeing, W., "Electrochemical machining of heat resistant alloys", *Annals CIRP*, No. 1, Vol. 21, p. 43, 1972.

40. Kawafune, K., *et al.*, "Study of shapes processed by ECM", *CIRP Annals*, Vol. 16, p. 345, 1968.

41. Merchant, M.E., "Newer methods for the precision working of metals-research and present status", *Advances in Machine Tool Design and Research*, Oxford, Pergamon Press, Ltd., p. 93, 1962.

42. Anon., "Electrochemical machining rates", *Tool and Manf. Engr.* Vol. 58, p. 82, June 1976.

43. Wilkinson, B., and P. Warburton, "Electrochemical machining", *Proc. Conference on Machinability*, Iron and Steel Institute, London, 1965.

44. Maleka, A.H., "The applications of electrochemical machining", *Proc. Conf. on Machinability*, Iron and Steel Institute, London, 1965.

45. Molloy, J., "Electrochemical machining", *Machinery*, No. 2765, Vol. 107, p. 1027, November 1965.

46. Laboda, M.A. and M.L. McMillan, "New electrolyte for electrochemical machining", *Electrochemical Technology*, Vol. 5, p. 340, August 1967.

47. DeBarr, A.R. and D.A. Oliver, *"Electrochemical Machining*, Mcdonald, London, 1968.

48. Faust, C.L., *Fundamentals of Electrochemical Machining*, The Electrochemical Society, U.S.A., 1971.

49. Bannard, H., "Fine hole drilling using electrochemical machining", *Proc. 19th International Machine Tool Design & Research Conf.*, The Macmillan Press, U.K., p. 503, 1979.

50. Larsson, C.N. and K. Muzaffarudin, "Electrochemical effects on shape reproduction in electrocehmical machining", *Proc. 19th International Machine Tool Design & Research Conf.*, The Macmillan Press, U.K., p. 533, 1979.

51. Tipton, H., "The Calculations of Tool Shapes for Electrochemical Machining", *Fundamentals of Electrochemical Machining*, Electrochemical Society, Princeton, 1971.

52. Cole, R.R., "Basic research in electrochemical machining – present status and future directions", *International Journal of Production Research*, Vol. 4, p. 75, 1965.

53. Kubeth, H. and H. Heitmann, "Electrochemisches senkenein neus abtrangendes verfahren zur metall bearlbeituing", Industrie Anzeiger, Vol. 85, p. 46, 1963.

54. Molloy, J.K., "Comparison between electrochemical and conventional methods of machining complex shapes", *Machinery*, Vol. 108, p. 13, 1966.

55. Cuthbertson, J.W. and T.S. Turner, "Electrochemical machining – A study of the effect of some variables", *Prod. Engr.* Vol. 45, p. 270, 1966.

56. Zimmel, L.J., *An Analysis of Effects of ECM on Fatigue of 403 Stainless Steel*, Paper presented at National Aeronautics and Space Engineering Meeting, October 1964.

57. Brandi, R.R., "Basics of electrochemical grinding", *American Machinist*, No. 9, Vol. 118, p. 45, 1974.

58. Dietz, H. *et al.*, "Electrochemical machining: calculation of side gap w.r.t. hydrogen evolution", *Annals CIRP*, No. 1, Vol. 23, p. 45, 1974.

59. Koenig, W. and H. Degenhardt, "Electrochemical machining of heat resistant alloys", *Annals CIRP*, No. 1, Vol. 21, p. 43, 1972.

60. Landolt, "Mechanical aspects of electrochemical machining of metals" (in German), *Chem. Ing. Tech.*, No. 4, Vol. 45, p. 188, 1973.

61. Loutrel, S.P.and N.H. Cook, "High rate electrochemical machining", *ASME* Paper n-73-Prod-3 for meeting (May 1973).

62. Meleka, A.H., "ECM: area of application, possibilities and limitations" (in German), *TZ Prakt Metallbearb*, No. 3, Vol. 69, p. 71, 1975.

63. Venkatakrishnaiah, C.O. *et al.*, "Some aspects of ECM", *Journal of Inst. of Engineers* (India), pt. ME4, Vol. 55, p. 180, March 1975.

64. Thorpe, J.F., "Operational characteristics of ECM", *Fertigung* (in German), No. 2, Vol. 6, p. 47, 1975.

65. Kops, L., "Effect of the workpiece structure on ECM", *Fertigung* (in German), No. 2, Vol. 6, p. 53, 1975.

66. Marty, C., "Fundamental principles of chemical machining", *Mecanique-Materiaux-Electricite*, No. 308, Vol. 58, p. 15, August-September 1975.

Electrochemical Grinding

67. Trott, L.L., "ECG finishes micro-miniature parts", *Tool and Manufacturing Engineer*, Vol. 52, p. 81, January 1964.

68. Storey, O.W., "Electrolytic grinding or machining of metals", *Journal of Electrochemical Society*, Vol. 100, p. 125, May 1953.

69. Cole, R.R., "An experimental investigation of electrolytic grinding process", *ASME Journal of Engineering for Industry*, Vol. 83, p. 194, May 1961.

70. Shan, H.S., "A theoretical analysis of electrolytic grinding process", *Microtecnic*, Vol. 25, p. 472, 1971.

71. Reinhart, H., and W. Grunwald, "Electrolytic stock removal of sintered carbide with diamond grinding wheels", *Industrial Diamond Review*, No. 266, Vol. 23, p. 19, January 1963.

72. Kaczmarek, J. and T. Zachwieja, "Investigations on material removal rate by electrochemical grinding of cutting tool materials in dependence on the properties of the grinding wheel", *International Journal of MTDR*, Vol. 6, p. 1, 1966.

73. Bass, M.M., "Electrolytic grinding—production process now", *Machine Tool Blue Book*, Vol. 60, p. 116, November 1965.

74. Scholz, E., "Electrochemical profile grinding", *Industrie Anzeiger*, Vol. 88, p. 1522, August 1966.

75. Levin, Z.D., "Electrolytic form grinding of cemented carbide tools", *Machines and Tooling*, Vol. 34, p. 34, 1963.

76. Sata, K., "Electrolytic grinding applied to special high speed steel", *Journal of Japan Society of Precision Engineering*, Vol. 30, p. 462, 1964.

77. Kubota, M., "Study on electrolytic abrasive machining (First report): Electrolytic lapping of cemented carbide", *Journal of Japan Society of Precision Engineering*, Vol. 31, p. 234, 1965.

78. Schadach, P., "Planning and Operation of electrolytic grinding plant", *Werkstatt-stechnik*, Vol. 45, p. 426, 1965.

79. Pahlitzsch, G., "Stosslaepp-elysieren-ein neus elektro-physikalisches abrogverfahren", *CIRP Annalen*, Vol. 11, p. 67, 1962-63.

80. Koldos, E., "A study of electrochemical grinding of small diameter specimens made of tungsten carbide", *Periodica Polytechnica*, Vol. 18, p. 21, 1974.

Electrolytic Deburring

81. Wick, C.H., "ECM: a new tool for deburring", *Machinery* (NY) Vol. 71, p. 143, March 1965.

82. Schoffler, G., "Electrolytic deburring—a new method of rationalizing production", *Werkstatt und Betrieb*, Vol. 96, p. 865, December 1963.

83. Williams, L.A., "Connecting rods deburred via ECM", *Steel*, Vol. 157, p. 49, December 1965.

84. Boyle, S.C., "Electrochemical deburring of aluminium components", *Light Metals*, Vol. 27, p. 48, May 1964.

Electrochemical Honing

85. Eshelman, R.H., "Electrochemical honing reports ready for production jobs", *Iron Age*, p. 124, August 1963.
86. "Honing gets electrolytic help", *Iron Age*, p. 106, April 1965.
87. "Electrochemical honing nearing production stage", *Steel*, Vol. 157, p. 44, August 1965.
88. "Electrochemical honing—four times faster", *Machinery* (NY), Vol. 72, p. 119, October 1965; also *Grinding and Finishing*, Vol. 11, p. 32, October 1965.
89. "Tooling for electrochemical honing", *Tool and Manufacturing Engineer*, p. 68, October 1965.

Chemical Machining

90. Astrop, A.W., "Chemical milling challenges presswork", *Machinery and Production Engineering*, No. 3154, Vol. 122, p. 566, 1973.

Electric Discharge Machining

91. Smith, G.V., "Spark machining—fundamentals and techniques", *Journal of the British Institution of Radio Engineers*, Vol. 22, p. 409, November 1961.
92. Alden, C.R., "Electrospark machining", *Mechanical Engineering*, No. 9, Vol. 75, p. 701, September 1953.
93. Sandford, J., "Cutting with spark", *American Machinist*, No. 14, Vol. 110, p. 77, July 1966.
94. Chevallier, R., "Spark machining of forging dies", *Metal Treatment and Drop Forging*, No. 151, Vol. 25, pp. 135, 144, April 1958.
95. Grodzinski, P., "Unorthodox methods of machining hard materials", *Metallurgica*, No. 279, Vol. 47, p. 34, January 1953.
96. Rudorff, D.W., "Spark machining and its development", *Metal Treatment and Drop Forging*, No. 172, Vol. 27, pp. 4, 11, January 1960.
97. Rudorff, D.W., "Spark machining equipment", *Metal Treatment and Drop Forging*, No. 186, Vol. 28, p. 120, March 1961.
98. Rudorff, D.W., "Principles and applications of spark machining", *Institution of Mechanical Engineers Proc.*, No. 14, Vol. 171, p. 495, 1957.
99. Springborn, R.K., "Nontraditional machining processes", *ASTME*, Michigan, 1967.
100. Barash, M.M., "Electric machining of metals", *Modern Workshop Technology*, Part II, Editor H. Wright Baker, Second edition, p. 290, 1960.
101. Williams, E.M. and T.O. Hokenberry, "Dynamic evaluation of events accompanying low voltage discharges employed in EDM", *IEEE Transactions on Industry and General Applications*, No. 4, Vol. IGA-3, p. 302, July-August 1967.
102. Karbacher, E.J., W.A. Haggerty, C.R. Allison and M.F. Devise, "Electrical methods of machining", *International Research in Productions Engineering*, Proc. of the International Production Engineering Research Conference, *ASME*, p. 232, September 1963.
103. Divers, S.V., "Spark machining as an aid to production", *The Production Engineer*, No. 3, Vol. 41, p. 140, March 1962.
104. Benard, C., "The present state of the development of spark-erosion generators", *Machinery*, No. 2577, Vol. 100, p. 772, April 1961.
105. Farley, F.G., "Characteristics of spark erosion circuits", *Metal Treatment and Drop Forging*, No. 170, Vol. 26, p. 405, November 1959.
106. Blake, L.R., "High power spark erosion machine", *The Engineer*, No. 5159, Vol. 199, p. 222; February 1955.

107. Anon., "Electric spark machining", *Machinery* (London), No. 2069, Vol. 81, p. 57, July 1952.

108. Anon., "The linderode spark erosion process", *Machinery* (London), No. 2597, Vol. 101, p. 427, August 1962.

109. Lawrance, A.J., "The application of spark erosion machining", *The Institution of Production Engineer*, No. 11, Vol. 37, p. 194, November 1958.

110. Adcock, J.L., "Electrode materials", *Metal Treatment and Drop Forging*, No. 174, Vol. 27, p. 103, March 1960.

111. Rhodes, G.H.C., "Practical aspects of cavity sinking", *Metal Treatment and Drop Forging*, Vol. 27, p. 151, 1960.

112. Ullmann, W. and T.G. Tranbe, "Practical and economical problems of machining cavities by spark erosion", *Metal Treatment and Drop Forging*, Vol. 27, p. 493, 1960.

113. Herridge, F.W., "Spark machining for tool production and repetition work", *Machinery* (London), No. 2880, Vol. 112, p. 148, 1968.

114. Anon., "Developments in the sparcatron spark-machining process", *Machinery* (London), No. 1281, Vol. 85, p. 488, September 1954.

115. Anon., "Czechoslovakian spark machining equipment", *Machinery* (London), Vol. 89, p. 229, July 1956.

116. Anon., "Wickman type W/DM erodomatic machine", *Machinery* (London) No. 2443, No. 2280, Vol. 95, pp. 492, 499, September 1959.

117. Anon., "Burton Griffiths machine tool demonstration", *Machinery*, (London), No. 2425, Vol. 24, p. 1038, May 1959.

118. Anon., "Sparcatron MKV spark erosion equipment", *Machinery* (London), Vol. 97, p. 344, August 1960.

119. Anon., "Elox spark machining equipment", *Machinery* (London), No. 2477, Vol. 96, p. 985, May 1960.

120. Davis, M.F., "A symposium on comprehensive investigations in EDM", Creative Manufacturing Seminars, *ASTME*, Michigan, USA.

121. Davis, M.F., "Elements, applications and economics of electro-discharge machining", Paper No. 6S5, *ASTME*, Michigan, USA.

122. Chowdhary, S., A.K. Das and A. Bhattacharya, "Determination of critical resistance in spark erosion machining for various electrodes in a given dielectric", *Proc. 2nd AIMTDR Conference*, 1968.

123. Kurafugi, H., "Development of research and application on spark erosion and electrolytic machining", *14th CIRP*, General Assembly, September 1964.

124. Lazarenko, B.R., "The present state of development of electro spark machining of conductive materials abroad", *Electrospark Machining of Metals*, Vol. 2.

125. Lazarenko, B.R. and N.I. Lazareako, "Technological characteristics of electro-spark machining of current conducting materials", *Electrospark Machining of Metals*, Vol. 2. New York Consultants Bureau.

126. Bhattacharya, A., *New Technology*, Inst. of Engineers (India), 1973.

127. Crookall, J.R. and B.C. Khor, Electrodischarged machined surfaces, *Proc. 15th International MTDR Conference*, 1974.

128. Crookall, J.R. and P.W. Lee, "Some effects of debris concentration on erosion and electrode wear in EDM", *Proc. 15th International MTDR Conference*, 1974.

129. Hughes, D.G. and J.L. Sheldon, "Electrical Machining", *Proc. 14th International MTDR Conference*, 1973.

130. Taniguchi, N., N. Kinoshita and M. Fu Kui, "Optimum form of the current impulse in EDM", *Annals of CIRP*, Vol. 20/1, 1971.

131. Heuvelman, C.J., H.J.A. Horsten and P.C. Veenstra, "An introductory investigation of the breakdown mechanism in EDM", *Annals of CIRP*, Vol. 20/1, 1971.

132. Rasch, F.O., "Micro-crack distribution on surfaces generated by ECM and EDM", *Annals of CIRP*, Vol. 20/1, 1971.

133. Bhattacharya, S.K. and M. Kettle, "Some observations on spark erosion machining", *Production Engineer*, No. 9, Vol. 51, p. 305, September 1972.
134. Kettle, M., *Electrical Discharge Machining*, Report 71, Department of Engineering Production, University of Birmingham.
135. *Introduction to EDM*, Charmiles-Geneva, 1968.
136. Livshits, A.L., *Electro-erosion Machining of Metals*, edited by R.S. Bennet, Butterworths, London, 1960.
137. Smith, G.V., *Electrical Methods on Machining and Forming*, Proc. of Conference on Institute of Electrical Engineering, London, p. 48, 1967.
138. Kok, A.J., *Conference on Electronic Processes in Dielectric Liquids* (Durham), 1963.
139. *Electrical Machining Handbook*, Cincinnati Milacron Inc., Ohio, USA.
140. *Metal Working Production Breakthrough in Qualifying EDM*, p. 77, June 1967.
141. "Adaptive control optimises EDM", *Metal Working Production*, p. 32, May 1971.
142. Sandford, J.E., "Production EDM", *Iron Age Review*, November 1968.
143. "EDM slashes airline maintenance costs", *Metal Working Production*, p. 61, April 1969.
144. Jeswani, M.L. and N. Ramaswamy, "Technical characteristics of electrospark machining processes", *Proc. IV AIMTDR Conference*, 1970.
145. Dave, B.J., Sevro controlled electrical discharge machines, *Proc. IV AIMTDR Conference*, 1970.
146. Heitman, H. and M. Adithan, "How to select and use a spark erosion machine", *Journal of Inst. of Engineers (India)*, January 1970.
147. Pal, M.N., P.K. Mishra and A. Bhattacharya, "Optimization of circuit parameters of a relaxation circuit of EDM", *Proc. IV AIMTDR Conference*, 1970.
148. Kurafuzi, H. and Suda Ko, "Study of Electrical Discharge Machining", *Proc. First AIMTDR Conference*, 1967.
149. Barash, M., "Some properties of spark eroded surfaces", *Microtecnic*, Vol. 13, p. 1. 1959.
150. Barash, M., "Electric spark machining", *International Journal of MTDR*, Vol. 2, p. 281, 1962.
151. Mironoff, N., "Electro-erosive metal working, its physical fundamentals and industrial applications", *Microtecnic*, Vol. 19, p. 149, 1974.
152. Crokall, J.R., "EDM The state of art", *Annals CIRP*, No. 2, Vol. 20, p. 113, 1971.
153. Greene, J.E., "Electro-erosion of metal surfaces", *Metallurgical Transactions*, No. 3, Vol. 5, p. 695, 1974.
154. Koenig, W. *et al.*, "Progress in electrochemical and electrospark processing techniques", *Giesserei* (in German), No. 3, Vol. 61, p. 51, 1974.
155. Gschanaidner, R. and G. Wagner, "Spark erosion machining of cooling air bores in turbine blades", *Werkstatt and Betrieb*, No. 3, Vol. 107, p. 155, 1974.
156. Sandford, J.E., "EDM spark in as high production tool", *Iron Age*, No. 18, Vol. 211, p. 55, 1973.
157. Ramaswamy, R. and S.L. Raj, "Study of wear and surface finish during spark erosion machining of high speed tool steel", *Wear*, No. 2, Vol. 24, p. 153, 1973.
158. Krishnan, A., "Improved relaxation generator for spark erosion machining", *Microtecnic*, No. 4, Vol. 27, p. 180, 1973.
159. Crookall, J.R., "Basic analysis of pulse trains in EDM", *International Journal of Machine Tool Design & Research*, No. 3, Vol. 13, p. 199, 1973.
160. "Application of electrospark machinery methods in the manufacture" (in German), *Industrie Anzeiger*, No. 9, Vol. 96, p. 2061, November 1974.
161. Wack, L.C., "Machinability and surface integrity—EDM, ECM", *Proc. Metal Process and Fabrication Seminar*, 1974.
162. Koenig, W. *et al.*, "Material removal and energy distribution in EDM", *Annals CIRP*, No. 1, Vol. 24, p. 95, 1975.

163. Van Dijck, "Theoretical and experimental study of the main parameters governing the EDM process", *Proceeding Group pour L'Av de la Mec Ind.*, [Spring meeting (Methodes d 'Usinage Nontraditionelles), Paris, 1974.

164. Van Dijck, "Metal removal and surface layer in EDM", *Proc. International Conference on Production Engineering*, Tokyo, Japan, August 1974.

Plasma Arc Machining

165. Dyos, G.T., "Effects of vibration and pulsation on metal removal by a plasma torch", *Journal of Heat Transfer, Trans. ASME*, Vol. 95, May 1973.

166. Mushenko, N.Ya, *et al.*, "The mechanised oxygen cutting of risers of heavy steel castings", *Weld. Prod.* No. 11, Vol. 24, p. 41, 1977.

167. Hebble, C.M. Jr., "Plasma cutting—an exotic process comes into its own", *Welding Engineering*, No. 3, Vol. 58, 1973.

168. Karly, Vilgem Zavod Za Varjenje, "Economic comparison on plasma arc cutting using different gases as gas mixtures", *Schweisstechnic*, No. 6, Vol. 23, p. 249, 1973.

169. Nachman, M., "Few considerations about the designing of plasma arc torches", *Rev. Int. Hautes Temp. Refract*, No. 2, Vol. 10, p. 65, 1973.

170. Kharitonov, E.P., "Magnetic control of plasma arc during the cutting of metals", *Weld. Prod.*, No. 12, Vol. 19, p. 73, December 1972.

171. Osadin, B.A. and N.V. Rusakov, "Consideration of an erosion plasma", *Soviet Physics, Technical Physics*, No. 2, Vol. 19, August 1974.

172. Hoeffer, El., "Plasma arc cuts bottlenecks", *American Machinist*, No. 21, Vol. 117, 1973.

173. Vasileev, K.V. *et. al.*, "Evaluation of ways of increasing the productivity of machine thermal cutting", *Weld. Prod.*, No. 3, Vol. 21, p. 65, 1974.

174. Budnik, N.M., "Plasma spraying of a protective coating from aluminia on refractory materials", *Weld. Prod.*, No. 12, Vol. 20, p. 26, December 1973.

175. Ingham, H.S. Jr. and A.J. Fabel, "Comparisons of plasma flame spray gases", *Weld. Journal*, No. 2, Vol. 54, p. 101, 1975.

176. Metcalf, J.C. *et al.*, "Heat transfer in plasma arc welding", *Weld. Journal*, No. 3, Vol. 54, p. 99, 1975.

177. Vishnovetsk, Y. M. *et al.*, "Mechanical plasma arc cutting of blanks with a curved surface", *Weld. Prod.*, No. 1, Vol. 21, p. 76, 1974.

178. Mudrov, M.V. *et al.*, "Equipment for gas cutting of circular component from rolled sheet", *Weld. Prod.*, No. 12, Vol. 20, p. 63, 1973.

179. Gaponenks, L.N. *et al.*, "Section for the air plasma cutting of high alloy steel", *Weld. Prod.*, No. 11, Vol. 20, p. 61, 1973.

180. Tbilisi Division of VNIIESO, "Plasma welding equipment", *Weld. Prod.*, No. 11, Vol. 20, p. 44, 1973.

181. Briskman, A.N. *et al.*, "Plasma arc cutting of stainless steel in the manufacture of chemical machinery", *Weld. Prod.*, No. 6, Vol. 20, p. 84, 1973.

182. Beider, B.D. *et al.*, "Plasma arc cutting in sea water", *Weld. Prod.*, No. 6, Vol. 22, p. 86, 1973.

183. Beider, B.D. *et al.*, "Equipment OP R-6-3m for plasma arc cutting", *Weld., Prod.*, No. 9, Vol. 21, p. 79, 1974.

184. Stol'bov, *et al.*, "The plasma cutting of aluminium alloys using a three phase arc", *Weld. Prod.*, No. 2, Vol. 24, p. 36, 1977.

185. Shalimov, A.P., "Technological and energy characteristics of air plasma gouging", *Weld. Prod.*, No. 10, Vol. 23, p. 47, 1976.

186. Stepanov, V.V. and V.I. Nechaev, "On pressure of the plasma arc", *Weld. Prod.*, No. 11, Vol. 21, p. 6, 1974.

187. Otkidach, L.G. *et al.*, "The fluxless cutting of stainless steels", *Weld. Prod.*, No. 7, Vol. 21, p. 68, 1974.

188. Beider, B.D. *et al.*, "Determination of the parameters of a plasma arc in a stream of argon and nitrogen used for the cutting of metals", *Weld. Prod.*, No. 6, Vol. 21, p. 54, p. 75, 1974.

189. Larri, O. Ya, "Equipment for the air plasma cutting of metals", *Weld. Prod.*, No. 5, Vol. 21, p, 75, 1974.

190. Syrovatkin, A.A. *et al.*, "Optimization of the inner geometry of a plasmatron for the cutting of metals", *Weld. Prod.*, Vol. 21, p. 18, 1974.

191. "Seminar on electrical cutting of metals", *BARC* (India), 1972.

192. Shapiro, I.S. *et al.*, "The influence of parameters of a gas supply system on the operating characteristics of plasmatrons for metal cutting", *Weld. Prod.*, No. 7, Vol. 23, p. 24, 1976.

193. Cresswell, R.A., *Plasma Cutting*, The Welding Institute, Research Bulletin, 11 (8), p. 229, 1970.

194. Livinstein, M.A. *et al*, "Properties of plasma sprayed materials", *Welding Journal* (USA), No. 1, Vol. 40, p. 85, 1961.

195. Vasil'ev V.A. *et al.*, "Calculation of the voltage of a cutting plasma arc", *Weld. Prod.*, No. 8, Vol. 23, p. 17, 1976.

196. Kronic, N.A. *et al*, "Mechanised surfacing with ceramic roads", *Weld. Prod.*, No. 5, Vol. 23, p. 41, 1976.

197. Pilkin, E.I. *et al.*, "Special features of the oxyacetylene cutting of low carbon steel", *Weld. Prod.*, No. 10, Vol. 24, p. 45, 1977.

198. Rao, K.N. and G.J. Gururaja, "Some studies of the plasma cutting of aluminium plates", *Indian Welding Journal*, No. 3, Vol. 6, p. 80, July-September 1974.

Electron Beam Machining

199. Bas, E.B., "Materialbearbeiting mit electronen in hochvakuum", *Advances Vacuum Science Technology*, Vol. 2, p. 691, 1960.

200. Christy, R.W., "Formation of thin polymer films by electron bombardment", *Journal of Applied Physics*, Vol. 31, p. 1680, 1960.

201. Ennos, A.E., "The sources of electron induced contamination in kinetic vacuum systems", *British Journal of Applied Physics*, Vol. 5, p. 27, 1954.

202. Ennos, A.E., "The origin of specimen contamination in the electron microscope", *British Journal of Applied Physics*, Vol. 4, p. 101, 1953.

203. Yoshida, S., "Radioactive isotope study of the dissociation of barium oxide under electron bombardment", *Journal of Applied Physics*, Vol. 27, p. 497, 1956.

204. Baker, A.G. and W.C. Morris, "Decomposition of metallic films by electron impact decomposition of organometallic vapours", *Review Scientific Instruments*, Vol. 32, p. 458, 1961.

205. Anon., Some elementary considerations in design of electron beams for welding and heating, *Proc. 3rd Symposium on Processes*, Boston, p. 9, March 1961.

206. Anon., "Electron beam processing", *Aircraft Production*, No. 8, Vol. 22, p. 282, 1960.

207. Namba, S., *Applications of Electron Beams for Evaporation and Machining of Materials*, Proc. 4th Symposium on Electron Beam Technology, Boston, p. 304-23, March 1962.

208. Duhamel, R.F., *Practical Electron Beam Cutting and Milling Applications*, ASTME Creative Mfg Seminars—Tech. Paper SP63-29, p. 14, 1962.

209. Pahlitzsch, G., "Ursachen der thermischen beeinflussung des werkstueckes bei der elektronemstrahlbearbeitung", *CIRP Annals*, No. 2, Vol. 16, p. 93-201, 1968.

210. Kuper, G.I., Electron beam device for removal of dielectric and metallic materials", *VDI Z*, No. 5, Vol. 116, p. 401-08, 1974.

211. Miyazaki, T., "Pulse condition of electron beam drilling", *Bulletin Japan Society of Precision Engg.*, No. 1, Vol. 10, p. 35, 1976.

212. Miyazaki, T., "Characteristics of electron beam drilling of metals", *Bulletin Japan Society of Precision Engg.*, No. 2, Vol. 10, p. 71, 1976.

213. Von Dobeneck, "Material processing with electron beams—the process and its industrial applications", *TZ Prakt Metallbearb* (in German), No. 5, Vol. 6$ p. 156, 1975.

214. Skubich, J., "Laser and EBM in the production process", *Werkstatt and Betrieb* (in German) No. 7, Vol. 108, p. 425, 1975.

Laser Machining

215. Stovell, J.E. and B.F. Scott, "CO_2 laser machining", *Proc. Conf. on Electrical Methods of Machining, Forming and Coating*, Institution of Elec. Engineers, London, p. 7, 1970.

216. Babenko, V.P. and V.P. Tychinskii, "Gas-jet laser cutting", *Soviet Journal of Applied Physics*, No. 5, Vol. 2, p. 399, 1973.

217. Paek, U.C., "Thermal analysis of thin film micromachining with laser", *Journal of Applied Physics*, No. 5, Vol. 44, p. 2260, 1973.

218. Paek, U.C., "Thermal analysis of laser drilling processes", *IEEE Journal Quantum Electronics*, Vol. QE 7, No. 6, p. 277, 1971.

219. Engel, A., "Laser machining with modulated zone plates", *Applied Optics*, No. 2, Vol. 13, p. 269, February 1974.

220. Bastien, J., "Applications of laser to welding and machining", *Soudage Tech. Connexes* (in French), No. 11-12, Vol. 25, p. 437, November-December 1971.

221. Pietermatt, F.P., "Thermic applications of lasers", *Bulletin Scientifique de I'AIM*, No. 3, Vol. 84, p. 1, 1971.

222. Schmidt, A.O. *el al.*, "Laser features and applications to micromachining and microwelding", *Proc. 5th Int. MTDR Conference*, p. 565, 1964.

223. Anon., "Laser and its application in machining", *Mech. Miesiecznik Nauk-Tech.* (in Polish), No. 3, Vol. 43, p. 137, 1970.

224. Paszkowski, B. *et al.*, "Some problems of laser beam welding and drilling", *Electron Technology*, No. 1, Vol. 2, p. 175, 1969.

225. Whiteman, P., "Laser machining of metals", *Machinability*, Iron and Steel Institute Special Report, p. 160, October 1965.

226. Kevern, J., "Advanced lasers qualify as practical machine tools", *Product Engineering*, No. 19, Vol. 37, p. 37, September 1966.

227. Houldcraft, P.R., "The importance of laser for cutting and welding", *Welding and Metal Fabrication*, p. 42, February 1972.

228. Adams, C.M. and G.A. Hardway, "Fundamentals of laser beam machining and drilling", *Trans. I.E.E.E. Industrial and General Applications*, p. 90, March-April 1965.

229. Kobayashi, A., "Laser drilling of non metals", *Annals CIRP*, No. 1, Vol. 20, 1971.

230. Mori, M. and H. Komehara, "Study of ultrasonic laser machining", *Annals CIRP*, No. 1, Vol. 25, p. 121, 1976.

231. Brandt, G. *et al.*, "Cutting with a CO_2 laser", *Schwetssen Schneiden*, No. 2, Vol. 23, p. 56, 1971.

232. Farwer, A., "Laser cutting of aluminium", *Aluminium*, No. 9, Vol. 47, p. 562, 1971.

233. Bebenko, V.P. *et al.*, "Program-controlled laser unit for cutting of materials", *Soviet Journal of Quantum Electron.*, No. 1, Vol. 3, p. 82, 1973.

234. Fletcher, M.J., "High power laser cutting of metals", *Weld. Metal Fabrication*, No. 9, Vol. 41, p. 308, 1973.

235. Hoffmann, M., "Laser's future in metal work", *Weld. Metal Fabrication*, No. 1, Vol. 42, p. 6, 8, 1974.

236. Rauscher, G.K., "Machining with laser and electron beams—current status of technology", *Praktical Metallbearb*, No. 4, Vol. 68, p. 111, 1974.

237. Palitzsch, G. and U. Eisleben, "Machining of materials with laser beams" (in German), *Metall*, No. 10, Vol. 26, p. 101, 1972.

238. Droscha, H., "Laser beam as a cutting tool", *Werkstatt and Betrieb*, No. 6, Vol. 106, p. 365, 1973.

239. Bunting, K.A., "Toward a general theory of cutting : a relationship between the incident power density and the cut speed", *Journal of Heat Transfer, Trans. ASME*, No. 1, Vol. 97, p. 116, 1975.

240. Belforte, D., "*Precision Metal Cutting with Carbon Dioxide Lasers*, SME Technical Paper MR-75-460 for Meeting, May 1975.

Hot Machining

241. Okoshi, M., "Hot machining by electric current", *International Research in Production Engineering, ASME*, p. 264, 1963.

242. Coster, A.S., E.F. Hatz and E.T. Armstrong, "Machining of heated metals", *Transaction of the ASME*, Vol. 73, p. 35, January 1951.

243. Merchant, M.E., "Basic factors in hot machining", *Trans. of ASME*, Vol. 73, p. 761, August 1951.

244. Barrow, G., "The effect of hot machining by electric current on the mechanics of orthogonal cutting", *Proc. 8th Int. MTDR Conf.*, p. 795, 1967.

245. Anon., "PERA hot machining technique for turning applications", Machinery and Production Engineering. No. 3051, Vol. 118, p. 711, 1971.

Index